Intersections of Formal and Informal Science

Science learning that takes place between and at the intersections of formal and informal science environments has not been systematically reviewed to offer a comprehensive understanding of the existing knowledge base. Bringing together theory and research, this volume describes the various ways in which learning science in various settings has been conceptualized, as well as empirical evidence to illustrate how science learning in these settings can be supported.

Lucy Avraamidou is an Associate Professor of Science Education, University of Groningen, Netherlands.

Wolff-Michael Roth is Lansdowne Professor of Applied Cognitive Science, University of Victoria, Canada.

Routledge Research in Education

Intersections of Formal and Informal Science

Edited by
Lucy Avraamidou &
Wolff-Michael Roth

Routledge
Taylor & Francis Group

LONDON AND NEW YORK

First published 2016 by Routledge

2 Park Square, Milton Park, Abingdon, Oxfordshire OX14 4RN
711 Third Avenue, New York, NY 10017

Routledge is an imprint of the Taylor & Francis Group, an informa business

First issued in paperback 2017

Library of Congress Cataloguing-in-Publication Data
Names: Avraamidou, Lucy. | Roth, Wolff-Michael, 1953–
Title: Intersections of formal and informal science / by Lucy Avraamidou and Wolff-Michael Roth.
Description: New York : Routledge, 2016. | Series: Routledge research in education ; 165.
Identifiers: LCCN 2015043499 | ISBN 9781138951051
Subjects: LCSH: Science—Study and teaching.
Classification: LCC Q181 .A97 2016 | DDC 510.71—dc23
LC record available at http://lccn.loc.gov/2015043499

ISBN: 978-1-138-95105-1 (hbk)
ISBN: 978-0-8153-8184-6 (pbk)

Typeset in Sabon
by Apex CoVantage, LLC

Contents

Figures

Tables

Preface

In recent years, researchers and institutions around the world have shown interest in the learning that takes place in informal science environments (e.g., museums) or out-of-school settings (e.g., science camps) and which operates across a broad range of contexts and disciplines and reaches out to people of all ages. This interest is related to the fact that people are very familiar with their everyday worlds, where they encounter many phenomena that also have pertinence in the formal sciences. These encounters generally are informal and framed in everyday language, such as when a person marvels at the beauty of a sunrise; and even scientists will marvel at the movement of the sun, rather than at the scientifically correct rotation of the earth that produces the appearance of a moving sun. Science educators have come to realize that such everyday experiences, although the associated discourses may be antithetical to science, nevertheless constitute a foundation and condition for anything scientific to occur.

This goal of connecting science to its everyday counterpart is based upon several underlying principles. One of these principles is connecting to students' interests and experiences. Such interests and experiences, and the enthusiasm that often come with them, are foundational to the subsequent emergence of formal science—even though the former frequently are overturned. It is obvious that the overarching goal covers a wide and demanding range of knowledge, skills, and attitudes toward science that are developed not only in school but also in out-of-school settings. Unless there are opportunities for linking the informal and the formal aspects of science, the latter will be disconnected in the same way as theories are viewed as disconnected from praxis generally.

Extensions beyond the school come in various forms. Informal science environments, such as science museums, natural history museums, cultural and history museums, zoos, aquariums, botanical gardens, science centers, after-school programs, and everyday life settings such as the community and the family environment offer unique educational environments and provide exciting opportunities for learning. Learners may develop awareness, interest, motivation, social competencies, and practices across informal social settings, such as on trips to museums and zoos, in the home, and in activities with friends and community projects. Accordingly, the informal education community pursues a range of learning outcomes that are connected to the idea of lifelong, life-wide, and life-deep learning. This idea has been influential in efforts to develop a broader notion of learning, incorporating how people learn over their life course, across social settings, and in relation to prevailing cultural influences. Philosophers and educators have recognized for quite some time that everyday knowing is the foundation of any formal system that culturally or individually is developed subsequently.[1] Thus students acquire knowledge through visits with their families to aquariums, zoos, parks, museums, and, in general, through spending time outdoor, playing games, or being involved in any activity in the context of informal environments.

The purpose of this volume is to describe the various ways in which learning science in informal settings has been conceptualized and to summarize empirical evidence drawn from studies situated in various informal contexts (e.g., formal schooling and informal sector collaborations, teacher preparation and informal sector collaboration, teacher professional development in informal contexts) to illustrate the significant role of informal science environments (museums, botanical gardens, outdoor settings, science centers, natural history museums) in supporting learning and development of the public, students, and teachers. In doing so, this book volume contains both theoretical and empirical chapters that (a) provide important conceptualizations of the learning that takes place in informal science environments, (b) provide empirical evidence of various programs that support student learning, and (c) provide empirical evidence of various programs that support preservice and experienced teachers' learning and development. Essentially, this book volume aims to respond to an important question for science education reform: *How can symbiotic relationships between formal and informal science be formed?*

In part, the attempt to find answers to this question requires overcoming the epistemological divides that many theoretical frameworks create between learning in schools and learning in settings outside of schools. These are explored through the five chapters in the first section of the book that offer a set of theoretical perspectives on learning that takes place in informal settings. These are followed by six chapters providing empirical evidence of programs, approaches, and experiences that support student learning and are situated in a variety of informal contexts. These range from the use of mobile technologies, to youth programs in botanical gardens, partnerships with scientists and videoproduction programs. The last section of the book volume includes six chapters that provide evidence of various programs and approaches to supporting teacher learning and development through collaborations between universities and informal science institutions.

As a relatively new area of research, science learning that takes place between and at the interface of formal and informal educational settings—where the question as to the sense in which "informal" is to be understood is itself debated—has not been reviewed systemically to offer a comprehensive understanding of what is currently known. This book volume contributes to closing this gap by reviewing existing theoretical perspectives and empirical evidence related to different programs and various approaches to supporting student learning and teacher learning in various types of informal contexts situated within different parts of the world. This book volume not only provides fresh theoretical perspectives on learning but also offers recommendations about how to address challenges and overcome constraints in attempting to achieve greater complementarity between the formal and the informal science sectors. The book constitutes a basis for conversations on issues of reconceptualization of learning, school science, teacher preparation, accreditation and teacher licensure, university policies, financial resources, and institutional cultural differences between informal science institutions, schools, and universities. As such, it will be of interest to academics, graduate students, teachers, informal science education staff, policy makers, and curriculum designers.

Groningen, Netherlands
Victoria, BC, Canada
October 2015

NOTE

1 The literature on this recognition, which is opposed to the constructivist view of overcoming and eradicating misconceptions, has been reviewed and elaborated in Roth, W.-M. (2015). "Enracinement or the earth, the originary ark, does not move: On the phenomenological (historical and ontogenetic) origin of common and scientific sense and the genetic method of teaching (for) understanding." *Cultural Studies of Science Education*, 10, 469–494.

Prologue

INTERSECTIONS OF FORMAL AND INFORMAL SCIENCE

Science education research traditionally focused on assessing students' conceptions, cognitive development, and learning in schools. In the 1980s, John Falk, later joined by Lynn Dierking, became a leading advocate of science learning that occurs in different settings, often focusing on field trips and museums (e.g., Falk, Koran, & Dierking, 1986). Later, in the wake of an increasing number of studies in the social studies of science on the importance of citizen involvement and the appearance of psychological studies emphasizing the situated nature of knowing, other science educators began advocating student involvement in everyday activities that may involve science (e.g., McGinn & Roth, 1999). Over the past 15 years, as evidenced in Thomson Reuters' Web of Science database, an increasing number of science education research studies include the keyword "informal science." Although many different theoretical approaches—e.g., cultural-historical activity theory, practice theories, or ethnomethodology—point out the ordered and orderly way in which human beings participate in *all* activities, the term "informal" tends to be used to demarcate whatever does not happen in formal educational settings (school, college, or university). There is recognition today that participation in non-curricular activities may generate interest, builds on individuals' familiarity with the world, and, therefore, supports science learning. Moreover, even when initial engagement in science does not correspond to scientists' understanding of phenomena, any "naïve" understanding actually constitutes the very possibility of science (Husserl, 1976). In the process of learning and development, initially naïve understanding is both overturned and retained (is sedimented), constituting the very foundation of formal science such that the latter is rooted in the familiarity that people develop prior to and outside of any formal science teaching.

An Overview of the Field

As illustrated earlier, the idea of "informal science" is not new, although it nowadays features centrally in contemporary reform efforts and reports around the world. In attempting to trace the development of ideas about learning that takes place outside the formal school realm, it becomes evident that the U.S. National Science Foundation (NSF) was the first funding institution to support the role of community organizations, museums, and media as essential partners in the educational process. In 1984, NSF created the Division of Informal Science Education based on the recommendation of a seminal publication, *Educating Americans for the 21st Century: A Report to the American People and the National Science Board.* Another important milestone in the development of the field of informal science education was the 2006 OECD Programme for International Student Assessment (PISA) survey, which "demonstrated that exposure to science-related extracurricular activities has positive relationships not only to student performance, but also to students' attitudes toward learning and their belief in their own abilities" (Rennie, 2014, p. 120).

Since then, quite a few centers, programs, and organizations specializing in informal science have been developed around the world, and several publications examining various aspects of informal science have made a strong presence in science education. Specifically, efforts to link formal and informal science education are evident in the development of centers (e.g., Center for Informal Learning and Schools, Center for Advancement of Informal Science Education) and are illustrated in numerous funded projects (e.g., EU-FP7: Permanent European Resource Centre for Informal Learning) university programs (e.g., museum studies, science communication) conferences, special interests groups and strands (e.g., National Association for Research in Science Teaching Strand Science Learning in Informal Contexts), and journal publications (e.g., *Visitor Studies, Museum Studies, Journal of Science Communication, International Journal of Science Education, Part B*, and the *Science Education* Science Learning in Every Day Life section). Key publications and reports in North America and Europe point to the importance of the intersections of the formal and informal science learning sectors. Examples of such reports are *Learning Science in Informal Environments*, commissioned in 2009 by the U.S. National Research Council, and the *Science Beyond the Classroom: Review of Informal Science Learning*, commissioned by the Wellcome Trust in the UK. These reports emphasize the importance of paying attention to how young people learn outside the classroom, as part of their everyday lives, and establishing links between these different settings of their daily experience.

A milestone for informal science education in Europe has been the establishment of the European Network of Science Centres and Museums (ECSITE). ECSITE was founded in Brussels by 23 organizations from the scene of the new European science centers. Over the past 25 years, the association has grown considerably, and it now gathers more than 350 organizations committed to inspiring people with science and technology and enabling dialogue between science and society. Concurrently, a few EU work programs have addressed the critical role of informal science. For example, the most recent one, HORIZON 2020, has as a priority the building of capacities and developing innovative ways of connecting science to society with the aim of making science education and careers attractive to young people. Informal science education features centrally in this call, which reads as follows:

> Innovative formal and informal science education teaching and learn-ing is important in order to raise both young boys' and girls' aware-ness of the different aspects encompassing science and technology in today's society and to address the challenges faced by young people when pursuing careers in Science, Technology, Engineering and Math-ematics (STEM).
>
> (European Commission, HORIZON 2020)

Likewise, in North America, the importance of informal science education is also evident in reform recommendations and related reports. In 2003, the policy statement of the National Association for Research in Science Teaching Ad Hoc Committee on Informal Science Learning described the broad nature of learning to emphasize the value of learning that happens in out-of-school settings:

> Learning rarely, if ever, occurs and develops from a single experience. Rather, learning in general, and science learning in particular, is cumula-tive, emerging over time through myriad human experiences, including, but not limited to, experiences in museums, schools, while watching television, reading newspapers and books, conversing with friends and family, and increasingly frequently, through interactions with the Inter-net. The experiences children and adults have in these various situa-tions dynamically interact to influence the ways individuals construct scientific knowledge, attitudes, behaviors and understanding. In this view, learning is an organic, dynamic, never-ending, and quite holis-tic phenomenon of constructing personal meaning. This broad view of learning recognizes that much of what people come to know about the world, including the world of science content and process, derives from real world experiences within a diversity of appropriate physical and social contexts, motivated by an intrinsic desire to learn.
>
> (Dierking, Falk, Rennie, Anderson, & Ellenbogen, 2003, p. 109)

These views found their way into the recently published *Framework for K–12 Science Education* as the basis for the development of new standards

(National Research Council, 2012). The report summarizes the overarching goal for K–12 science education in this way:

> By the end of 12th grade all students have some appreciation of the beauty and wonder of science; possess sufficient knowledge of science and engineering to engage in public discussions on related issues; are careful consumers of scientific and technological information related to their everyday lives; are able to continue to learn about science outside school; and have the skills to enter careers of their choice, including (but not limited to) careers in science, engineering, and technology.
>
> (p. 1)

This goal is based upon several underlying principles. One of these is about making connections between science and students' interests and experiences. According to the report, "research suggests that personal interest, experience, and enthusiasm—critical to children's learning of science at school or in other settings—may also be linked to later educational and career choices" (p. 28). It is obvious that the aforementioned overarching goal covers a wide and demanding range of knowledge, skills, and attitudes toward science that are developed not only in school but also in out-of-school settings. As a matter of fact, the National Science Education Standards proposed, "the school science program must extend beyond the walls of the school to include the resources of the community" (NRC, 1996, p. 45).

Grounded within these recommendations, researchers have examined different aspects of informal science that can be summarized into the following research areas: informal learning environments, museum studies, visitor studies, out-of-school learning, field trips, teacher training, and professional development in informal science institutions; and there may certainly be other areas. Consequently, in the literature, different conceptualizations of informal science can be found. The term "informal learning environments" is often used to describe learning environments outside the traditional area of schools. Such learning develops from various experiences and is an organic, dynamic, never-ending, and holistic phenomenon of constructing personal meaning. Teaching in environments outside schools therefore can create opportunities for students to get involved in activities where learning will be guided by their own interests and needs.

Other researchers have used the term "out-of-school learning" to study learning that happens outside the formal school realm. The definition of out-of-school learning may include that (a) it occurs out of school, is self-motivated, and guided by learners' needs and interests, (b) it is strongly socioculturally mediated, and (c) it is a cumulative process involving connections and reinforcement between the varieties of learning experiences a person encounters in life. These domains have been explored in informal science education and specifically in museum studies in the past decades with an emphasis on the role of exhibits on learning, with the use of various models and theoretical frameworks. The contextual model of learning is one framework for studying the complexities of learning within free-choice settings, such as museums (Falk & Dierking, 2000). Drawing upon constructivist, cognitive, and sociocultural theories of learning, this model exemplifies the contextually driven processes of interactions between a person's personal, sociocultural, and physical contexts. The personal context represents an individual's personal history (e.g., prior knowledge and experiences), the sociocultural context refers to the fact that museum learning is socioculturally situated given that individuals are defined by their social relationships and culture, and the physical context refers to the museum space and its characteristics. Over the past two decades, a number of studies investigated these contexts and how they impact learning that takes place in museums.

In the early years, including the 1970s and 1980s, research was devoted to investigating the cognitive aspects of learning with the use of quantitative research methods in the context of museums. A historical review of how museums and science centers have developed over the years illustrates how these settings have moved from hands-on displays and neglecting the processes of scientific practice and the nature of science to more critical exhibitions that consider socio-scientific issues from a variety of perspectives and address issues associated with the nature of science (Pedretti, 2002). Thus "science centers and museums are increasingly positioning themselves as socially valuable resources for conveying information to the public about science and technology and its social implications" (p. 34). Studies of museum learning research mostly look at exhibit-related experiences and points to the importance of:

- Structuring scientific knowledge: individuals can be shown to make conceptual gains from museum visits.
- The role of social interaction: family members engage in a variety of strategies that encourage explanation and understanding.
- Mediating experiences and devices: students' class visits to museums can relate to the depth and retention of visit content.
- Institutional meaning: individual' s identity as a learner is shaped by the cultural institutions he or she comes in contact with as well as by day-to-day moments in life.

(Martin, 2004, S73–74)

Over the past decade, a shift in the research could be observed to an examination of the social and affective domains of learning in museum settings and other out-of-school settings, such as community settings, family environments, science festivals, field studies, and scientists' laboratories (e. g., Avraamidou, 2013). The findings of such studies provide compelling evidence about the role of museums and out-of-school learning on supporting student learning and engagement with science. Section II of this book aims to add to this literature as it includes a set of chapters that offer examples of programs and research that examines science learning within a diverse set of science settings.

In the past few years, another trend is evident in the field of informal science education, which is illustrated through partnerships between universities and informal science institutions as part of teacher preparation programs as well as teacher professional development programs. Engagement in informal science education activities supports the development of teachers' science content knowledge, teachers' self-efficacy, and the development of positive attitudes toward science and science teaching, and the development of their identities. Ample research evidence shows that informal science environments

- Offer motivating structures for learning to teach and provide opportunities to practice science teaching in "safe" environments;
- Offer opportunities for learning to teach science through inquiry-based activities in environments that are rich in resources;
- Offer unique opportunities for developing content and pedagogical knowledge for science teaching; and,
- Can support teachers in developing understandings about the nature of science, the relationship of science to society, scientific inquiry and the work of scientists

(Avraamidou, 2014, p. 835)

The chapters of Section III fall within this research area and aim to contribute to existing knowledge about the significant role of informal science institutions to teacher preparation and teacher professional development.

The present book builds on and extends those studies. In addition to the many studies exhibiting the forms of learning that occur when individuals engage in science outside of formal school settings, although these often are formal in other ways (e.g., science museums, guided walks), we also come to know when public engagement with science does not have the intended effects and leads to the marginalization of people (e.g., Dawson, chapter 7). With the expansion of the field, one may not be surprised to hear also critical voices (e.g., Achiam, chapter 3; Dillon, chapter 5) and considerable rethinking of the very notions of learning, participation, and the role of settings not specifically organized for science (learning, teaching) (e.g., Roth, chapter 1). The present book therefore also offers ways to go beyond what we have come to know as "informal science" until now.

What to Expect from this Book

Throughout this book, the contributing authors articulate how and where people learn science, and it looks at the ways in which the *how* and the *where* intersect to provide opportunities for engagement with science. Many if not all contributors take the position that a democratic future depends on whether *all* people are offered exciting opportunities to engage with science and to become familiar with, and can talk about, how the world works. As such, we use the words "formal" and "informal" science in the title as a point of departure for attempting to respond to critical questions in science education: *What science? Where is science? Whose science? Science for whom? Who can do science? Will anybody do science? Will everybody do science?*

Historically, researchers have used different terms to refer to *informal science*, such as informal science learning, informal science environments or contexts, informal science institutions, outdoor learning, everyday learning, lifelong learning, and others. Generally, the term "informal science learning" has been used in the literature to refer to the learning that happens outside the formal school realm (Avraamidou, 2014). But, for example, aviation pilots also have very formal learning opportunities during their two annual training and assessment cycles, where, while debriefing their simulator performances with the flight examiner, they come to learn to fly in situations that they rarely if ever encounter in their normal routine (e.g., Roth, 2015). In disagreement with this view, we make neither an epistemological nor an ontological distinction between "learning" and "informal learning." Instead, we take the position that learning is learning no matter where it takes place. As stated earlier, some researchers have used the terms "informal learning environments" and "formal learning environments" to make a distinction based on the context where learning takes place. In our view, this is also problematic given that informal contexts, such as museums could, depending on how they are used, be formal and traditional learning contexts. Likewise, formal school environments could serve as places where opportunities for unorganized, real-world, informal science experiences are offered.

In attempting to move beyond the dichotomy of *space* and the *kind* of learning, we chose the title *Intersections of Formal and Informal Science*. This leaves open the conceptualization and definition of "formal" and "informal" science. As such, this book volume includes 17 chapters, whose authors offer a diverse set of conceptualizations, a variety of theoretical perspectives, and empirical evidence of successful intersections of formal and informal science within various geographical and cultural contexts and for a variety of age groups. The book consists of three sections: (a) Theoretical Perspectives; (b) Learning Science in Diverse Settings; and, (c) Universities and Informal Science Intersections. The first section consists of five chapters that offer a set of contemporary theoretical perspectives that challenge existing ideas about informal science and offer fresh conceptualizations of science learning that takes place in various contexts. The section that follows, includes six chapters that provide concrete examples of how people of different ages and backgrounds engage (or not) with science in a variety of places and how different kinds of approaches, programs, curricula, tools, and devices mediate or restrict their learning of and about science. The third section includes six chapters that offer concrete examples of intersections of informal science programs with universities either for teacher preparation or teacher professional development in a variety of cultural contexts. Each section is preceded by an introduction, which also includes a brief overview of the contents of the chapters.

The authors of the chapters are researchers in science education and informal science educators and practitioners. As such, the book volume offers a space where the voices of formal and informal sectors intersect in an attempt to bridge the two worlds and to provide multifaceted perspectives on how and where people engage (or disengage) with science. The rationale for exploring the intersections of formal and informal science is based on compelling research evidence that suggests that such bridging (a) increases student motivation for learning, (b) expands student conceptions of learning and knowledge, and (c) develops new student skills and abilities.

As stated earlier, we use the words "formal" and "informal" science as a point of departure for exploring the intersections between the infinite places where science takes place and the variety of ways that people engage with science. However, what we are really aiming at achieving is an exploration of the theoretical underpinnings and empirical knowledge that will allow us to move *beyond* the "formal" and the "informal" to conceptualize science learning in its totality, beyond dichotomies—to conceive the continuity of science learning across situations, time, and contexts (Roth, 2016). The challenge for science educators includes that it may not always be apparent initially that science is or will be involved in some contentious issue. Thus, for example, when the inhabitants of a small part of a Canadian municipality want to be connected to the water system that already supplies everybody else, it may not be evident that this is an issue of science. However, as soon as the health authority sends its scientists to investigate water quality and the politicians who are against building the water main hire other scientists to produce a divergent scientific report, then science has come to the foreground (e.g., Roth, 2008). Perhaps unfortunately, the "anecdotal" knowledge residents have accumulated about the water over four decades may be shoved aside in favor of this or that scientific study. Here, understanding and fostering diverse and heterogeneous forms of participation may take more science educators from their offices into the field. Science then comes to be experienced at the intersection with other aspect of public life, including law, politics, ethics (ecojustice), engineering, and environmentalism.

In conclusion, therefore, we encourage our readers not to get stuck with environments where science teaching occurs in some organized form, which includes museums, field trips, and investigations for preservice teachers, (online) media productions, science columns of (online) news media, or after-school programs. Instead, we hope readers take the assembled texts as a starting point for searching out all those other places that are not initially intended to present something from the field of science. Instead, there is a

real need to understand science teaching | learning in social life more gener-
ally, such as when citizens participate in planting eelgrass to regenerate a
habitat rather than to teach | learn science or when people fight for access
to drugs prior to final approval, because they do not want to be denied help
because they are part of the placebo group (as it has happened in the AIDS
case). What and how do hikers learn when they collect samples to contribute
to the documentation of fauna and flora in an area slated for logging? What
and how do bird enthusiasts (who may not use and even refuse the term
"amateur ornithologist") learn when seeking and observing the animals of
their interest? What and how do backyard hobby gardeners learn, especially
when exchanging with others living in their neighborhoods? What and how
do individuals learn once they decide to purchase a bee colony? Science
education is much too important as to be studied in a fraction of organized
situations where it occurs.

Lucy Avraamidou & Wolff-Michael Roth

REFERENCES

Avraamidou, L. (2013). Superheroes and supervillains: Reconstructing the mad-scientist
    stereotype in school science. *Research in Science and Technological Education,
    31*, 90–115.
Avraamidou, L. (2014). Developing a reform-minded science teaching identity: The
    role of informal science environment. *Journal of Science Teacher Education, 25*,
    823–843.
Dierking, L. D., Falk, J. H., Rennie, L., Anderson, D., & Ellenbogen, K. (2003).
    Policy statement of the "informal science education" ad committee. *Journal of
    Research in Science Teaching, 40*, 108–111.
Falk, J. H., & Dierking, L. D. (2000). *Learning from museums: Visitor experiences
    and the making of meaning.* Walnut Creek, CA: AltaMira Press.
Falk, J. H., Koran, J. J., & Dierking, L. D. (1986). The things of science: Assessing
    the learning potential of science museums. *Science Education, 70*, 75–89.
Husserl, E. (1976). *Husserliana Band VI. Die Krisis der europäischen Wissenschaf-
    ten und die transzendentale Phänomenologie. Eine Einleitung in die phänome-
    nologische Philosophie* [Husserliana vol. 6: The crisis of the European sciences
    and transcendental phenomenology. An introduction to phenomenological phi-
    losophy]. The Hague, The Netherlands: Martinus Nijhoff.
Martin, L. (2004). An emerging research for studying informal learning and schools.
    Science Education, 88, S71–S82.
McGinn, M. K., & Roth, W.-M. (1999). Preparing students for competent scientific
    practice: Implications of recent research in science and technology studies. *Edu-
    cational Researcher, 28*(3), 14–24.
National Research Council. (1996). *National science education standards.* Washing-
    ton, DC: National Academy Press.

National Research Council. (2012). *A framework for K–12 science education: Practices, crosscutting concepts and core ideas*. Washington, DC: National Academy Press.

Pedretti, E. (2002). T. Kuhn Meets T. Rex: Critical conversations and new directions in science centres and science museums. *Studies in Science Education, 37*, 1–41.

Rennie, L. (2014). Learning science outside of school. In N. G. Lederman & S. Abell (Eds.), *Handbook of research on science education volume II* (pp. 120–144). New York, NY: Routledge.

Roth, W.-M. (2008). Constructing community health and safety. *Municipal Engineer, 161*, 83–92.

Roth, W.-M. (2015). The role of soci(et)al relations in a technology-rich teaching | learning setting: The case of professional development of airline pilots. *Learning, Culture and Social Interaction, 7*, 43–58.

Roth, W.-M. (2016). Becoming and belonging: From identity to experience as developmental category in science teaching and teacher education. In L. Avraamidou (Ed.), *Studying science teacher identity* (pp. 295–320). Rotterdam: Sense Publishers.

Part I

Theoretical Perspectives

Any science teacher will have heard this question: "Why do I have to learn science?" Many (science) teachers—and sometimes parents as well—might have found themselves saying that it is important to make a good living. But looking around ourselves, we see that people often do well in life even if they know little or no science. Even scientists themselves often know little outside of their limited domain, and, as a number of studies show, fail to provide correct interpretations of graphs from introductory courses of their own fields (Roth, 2012). Much of science education appears to be stuck in the ideology of basic facts that we *need to know* not only to be scientifically literate but also to count as citizens, to be educated (aesthetically enriched), and to be able to appreciate the intellectual coherence of the world (e.g., Hazen & Trefil, 1991). In a world where there are an exponentially growing number of scientific publications, which of the zillions of scientific facts should school-aged individuals learn? Some studies show that even the most educated scientists may find themselves at a loss when caught up in a situation— such as personal health—where science is involved (Roth, 2014).

An important shortcoming of current science education appears to be that it fails to look at the lives of individuals in their entireties to answer the question of how science, if at all, might have a role as one strand among many others (see Figure E.1 in the epilogue of this book). It is here that out-of-school opportunities for learning have an important role, including, for example, science museums and other formal-informal settings. Such settings, although often thought about in terms of (free-choice) learning and informal settings, are in fact formally organized settings with specific, science-related purposes. There are many other settings in which science learning occurs but they are not framed in the same way. For example, environmental groups—such as the marine and watershed ecology related groups that Roth and his graduate students researched—do not inherently focus on science. If they do, it always is in the context of other dimensions of human life, such as engineering, politics, economics, health, ethics, public consciousness, and personal experience. In any environmental issue, these and other strands of human life are involved. Given that the ecological advantage of the human race derived from division of labor, we therefore need to think about science in the context of division of labor and in terms of the requirements for people with very different forms and levels of expertise to work together to solve a particular issue (Roth, 2003). It turns out that in such a context, scientists often exhibit scientific illiteracy because they appear to be incapable of communicating and negotiating with others in the assessment of the relative contribution science could make, rather than insisting on a unilateral dominance of science over all other fields of knowing and experience. Once we accept viewing science as but a strand among strands in the context of the whole lives of persons, we have made a first step toward understanding the contributions science actually can make. Chapter 1 takes such a broad perspective, which allows us to escape from the ego-centrist, within-science perspective that dominates science education today.

In "Learning: From Transitive Construction to Intransitive Being," Wolff-Michael Roth articulates a theoretical approach that no longer privileges the subject's agential engagement with the object via some form of mediator but orients us to the inherent changes that occur because we live and act in a world. His overall concern is with lifelong and life-wide changes, rather than merely those occurring in a school science classroom or a science museum. Using a practical example of a curriculum in which he, together with resident teachers, offered students opportunities to participate in environmentalism, he develops an alternative view on learning and a number of theoretical concepts. He sketches the reasons for his claim that constructivism is a dead end because it focuses on the *transitive* relations between subject and object. Grounded in the work of the anthropologist T. Ingold and the philosophers G. Deleuze and F. Guattari, the theoretical approach outlined orients us to the *intransitivity of becoming*. In the course of our lives, we find ourselves in many different contexts where we participate in doing what everyone else is doing and, in the course, are changed. Each person, each artifact, indeed each molecule is associated with a line of becoming; and when such lines come together and intertwine (as in a strand of wool, see "Epilogue"), they shape each other in the process. The chapter introduces *leading activity*, a concept developed by the Russian psychologist L. S. Vygotsky. This is a form of activity in which we participate and incrementally change (learn) but all of a sudden find ourselves having *developed* a *qualitatively* different form of consciousness and practice. Such developmental steps, according to Vygotsky, are unpredictable, suddenly and often unnoticeably arising in and from leading activity. Roth also introduces *experience* (pereživanie) as a theoretical category and analytic unit of the subject-acting-in-its-environment.

One step (or part) in the endeavor to rethink science education is to better understand the ways in which people are changed when they visit informal science institutions, such as science museums and science centers. Chapters 2 through 4 focus on the ways in which such places might be conceptualized to afford opportunities for free-choice learning, learning science content, and equity.

In "Free-Choice Learning: What Does It Mean?," Laura Martin from the Arizona Science Center addresses the question of how constructs such as "free-choice learning" help us understand how learning happens in different environments. She is concerned with the question of who wants to know what and the "production context" of the knowledge we have about learning science in formal and informal settings. She suggests that at the juncture of research and practice are competing explanatory and methodological frameworks attempting to pin down what constitutes evidence of "learning" in different circumstances. None of these alone explain the learning trajectories of individuals in different settings and the complex histories they bring to any particular program that allows them choice. After sketching the idea of free choice and the learner outside schools, she sets her readers up with answers to a series of questions: "What is knowledge and what counts as knowing?," "What do we know from practice?," "What do we know from research?," and "What do we know from theory?" She then turns to a discussion of how these insights add up and to what end. Can we usefully explore "free-choice" learning in different settings from different vantage points or does there need to be one story about them?

In "Attention to Content: Some Lessons From School-Oriented Education Research" (chapter 3), Marianne Achiam and Jan Alexis Nielsen argue that the content of science, as part of out-of-school science education, has been neglected in theoretical frameworks developed to explain how learners in out-of-school settings interact with other learners, the environment, and their prior knowledge to make meaning during their visits to these settings. In this chapter, the authors describe how the two contemporary conceptual frameworks (i.e., the generic learning outcomes and the contextual model of learning) that have generally been used to examine how learning takes places in out-of-school environments may have two unintended effects: (a) they may cause science centers/museums to disregard the discipline-specific ways in which the scientific content is represented and experienced in their dissemination activities and (b) neglect their disciplinary interpretative responsibility toward their visitors. The authors offer a set of possible solutions drawn from research situated in formal science education and, specifically, placing emphasis on the science content, setting learning objectives when designing exhibits, and using various types of assessment/feedback. This discussion is framed around two examples of science exhibits: a hands-on exhibit in the Paleontology Lab at the Royal Belgian Institute of Natural Sciences in Brussels and a hands-on exhibit at Experimentarium in Copenhagen.

"The Museum of Pink: Retheorizing the Science Museum" (chapter 4) written by Bronwyn Bevan from the Exploratorium in San Francisco is concerned with how the science museum—working at the intersection of formal, informal, and non-formal science education—can contribute to equity work. For the science museum to do such work, it needs to be retheorized to broaden our view of where such work occurs and what it looks like. She begins with the supposition that science museums are significant cultural resources in most urban landscapes. Whereas some, such as natural history museums, may also house vast collections, most science museums trade on providing their audiences *experience* with science as a field, as a process, and as a human endeavor. The possibilities for experiences with science vary across audiences. All of these audiences come to the museum equipped with their own prior experiences with and perspectives on science, driven by some sense of purpose, and informed by larger cultural representations of science. Bevan uses the Exploratorium as a case to articulate what is currently being done to expand possibilities for learning when science museums adopt ecological perspectives on learning and human development. Rather than conceptualizing themselves as a destination point, museums can actively position themselves as key resources for their communities if and when they recognize and leverage the purposes and resources that their varied audiences are developing and exercising in other contexts. She introduces the concept of *tinkering*, a form of engagement where people operate (and learn) at the boundaries of their current understanding; it therefore shares some important aspects with the notion of leading activity (chapter 1).

In the science education literature, (science) museums frequently are listed among the places where "informal" science learning may occur. In a strong sense, however, science museums, as museums generally, are *not* informal places at all. They are institutions in which people with various backgrounds—that may, but do not have to include, education degrees—organize exhibits with specific science concepts, forms of engagement, and display designs in view. The formal-informal dichotomy may actually not be appropriate when used to characterize school-museum relations. It is in this direction that Justin Dillon argues in "Beyond Formal and Informal."

In chapter 5, "Beyond Formal and Informal," Dillon encourages readers to rethink the use of formal and informal as descriptors of both learning and learning contexts. This is not a new argument, although it continues to be necessary to make it as the term "informal learning" still occurs in policy documents and educational materials. This chapter takes a critical look at the relationship, theoretically and practically, between science in the classroom and science outside—whether that is on school grounds, in museums, or in homes. In some cases, learning beyond the classroom reinforces existing understanding of scientific concepts; in other cases, it extends and challenges it. More recently the notion of a blended pedagogy—taking the best of school and museum education has emerged. What this blended pedagogy might look like is discussed in a number of contexts. Finally, the possibility of a convergence between science and environmental education through ICT-supported citizen science is discussed with a focus on breaking down the notion of formal/informal science education.

Readers will notice that despite the different approaches the authors of this section take, there are some important currents crossing them, which may be emphasized in one chapter but only implicit in another. Thus, for example, the idea of free choice underlies all chapters but is foregrounded in Martin's chapter. Free choice is in fact one of the fundamental human conditions that comes with the division of labor mentioned earlier: It does not matter where and in which way we contribute to generalized societal conditions (e.g., being farmers producing grain or vegetables), we assure expansion and control over our own personal conditions. However we contribute in productive activity, we secure the resources required to meet our personal needs. The content of the choice is free—although not choice itself—and even if we do not choose, we make a choice. To engage directly with science or to engage with someone knowledgeable in science are but two of the many possibilities we have in life. But there needs to be the possibility of deciding whether to engage with science at all—lest we become inconsistent with the democratic ideals underlying free-choice learning.

REFERENCES

Hazen, R. M., & Trefil, J. (1991). *Science matters: Achieving scientific literacy*. New York, NY: Doubleday.

Nietzsche, F. (1954). *Werke Band 3* [Works vol. 3]. Munich, Germany: Hanser.

Roth, W.-M. (2003). Scientific literacy as an emergent feature of human practice. *Journal of Curriculum Studies, 35*, 9–24.

Roth, W.-M. (2012). Limits to general expertise: A study of in- and out-of-field graph interpretation. In C. A. Wilhelm (Ed.), *Encyclopedia of cognitive psychology* (pp. 311–348). Hauppauge, NY: Nova Science.

Roth, W.-M. (2014). Personal health—personalized science: A new driver for science education? *International Journal of Science Education, 36*, 1434–1456.

1 Learning

From Transitive Construction to Intransitive Becoming

Wolff-Michael Roth

> As individuals express their lives, so they are. What they are therefore coincides with their production, both with *what* they produce and with *how* they produce. What individuals are depends on the material conditions of their production.
>
> (Marx & Engels, 1978, p. 21)

This book is about learning science generally and about learning at the intersection of formal and informal educational settings. Underlying much of the current debate is a constructivist epistemology, which, in some situations, appropriates for itself a sociocultural discourse—without nevertheless taking up the fundamentally dialectical and Marxist underpinnings of the work of the Russian psychologists that have given rise to that discourse. A Marxist view on learning is very different than that often conceived for the purpose of schooling with its decided intention to *make* institutionally designated students repeat a fundamentally outdated stock of knowledge; and they do so without really being studious, dedicated, zealous, and eager as per the etymological origin of "student" (Lat. *studēre*). We may understand learning in very different ways when we turn our gaze to what happens in informal settings, where there is less or no concern for making people—through the forces of an educational system where there is no free choice over contents and processes of participation—in the image of current science. In this chapter, I articulate a radically different view of learning, which arises from an epistemology that conceives of change not in terms of the *transitive* relation between the individual subject and its intentional object but *intransitively*; that is, in a direction that is transversal to the former. The view developed here also is consistent with anthropological and poststructuralist (postmodern[1]) conceptions of change associated with productive human life. To ground this theoretical discourse, I draw on a concrete situation where I taught science differently—traditional science educators may say perhaps I did not teach it at all—together with resident teachers.

DETERRITORIALIZING SCHOOL SCIENCE

Societal-Historical Context

Over the course of my professional career as a science teacher, I became dissatisfied with pre-specified lesson structures where students do not participate in determining what or how to research. I changed these conditions so that toward the end of my full-time teaching, the students in my classes designed their own investigations, nevertheless addressing the topics from the official curriculum of the education authority (Province of Ontario). Students (a) designed experiments; (b) wrote plays, puppet shows, or comic strips to serve as media of teaching a topic to their audiences; (c) designed curricula for and taught curricula to younger students in the school; or (d) did a library-based study on advanced areas of the topic. They also participated in the assessment of the products distributed over peer, teacher, and self-evaluation.

Despite all the self-determination that these classes offered, I soon began to realize that there was still too much orientation toward traditional schooling and too little toward students as human beings, members of a society, spending some part of their life trajectory in the place called school. A sense emerged in me that schooling—the societal formation that pronounces learning as its object but in fact produces inequality of access and deleted the lives, works, and voices of real human beings—is the problem. I began to think about the deinstitutionalization of school science, which was to be reterritorialized through participation in the community much as this had been the case in the movement from psychiatric institutions to community living. Drawing conclusions from a strong view on situated cognition and drawing on existing forms of activism written about in the social studies of science—AIDS, auricular implants, or myopathy—I began recommending different forms of tasks that would allow science learning to occur at the intersection of formal and informal education: seeding green corridors with butterfly pupa, monitoring pollution levels, hatching salmon for target creeks, biodiversity sampling, bird counting, habitat creation, and gardening as suitable endeavors that would take science education out of schools and into the community. After a change of jobs and the associated relocation, I had the test of the viability of my thinking while teaching three, three-to-five-month units in the municipality where I had moved.

Participating in Community Activism

Over a two-year period, I repeatedly taught an entire unit of seventh-grade science in the middle school of my community together with resident teachers. What I offered instead was a science course realized through participation in environmentalism. Thus the participating youths designed and conducted their own investigations in the local watershed with the intent to report their findings at an open-house event organized each year by an environmentalist group. I thought of the participants no longer in a traditional way, institutionalized learners, but as active citizens who dedicate themselves to contributing something to the community through their actions. Indeed, I developed the irreducible category of *activism* to theorize change—in participants and community—that arises when humans are engaged in changing the world, rather than merely attempting to understand it, which is the most optimistic view of what schools might be able to achieve. Members of the local activist group, the parents, and First Nations elders participated with us by providing workshops, talks, and assisting participants in framing what they wanted to do and in driving them to the different places in the watershed.

The participants readily bought into this approach, especially after we read with them an article from the local newspaper in which the director of the environmental group called for community participation in promoting— working for or advocating—the environmental health of the watershed. Almost all youths immediately wanted to do something about the creek, thereby expressing a sense of environmental justice with respect to this part of their environment that has become for some a ditch for disposing of their industrial waste. More so than their parents and older village inhabitants, these young people sensed that something was wrong with the creek in their community and that they could and should do something about it. Their expressed wishes to do something about the creek became even stronger once the director of the environmental group gave them a presentation about how they have been doing. Other individuals from the group also came to class or participated with different groups when they were already doing projects within the watershed. For example, a water technician working for a local farm showed the youths where and how water levels were measured, how the temperature and oxygen levels were determined, and how riffles and riparian trees increased oxygen levels. A biologist introduced some groups to the use of a colorimeter for measuring the turbidity of creek water. Interested participant groups subsequently got to use the instrument to conduct their own measurements as per the design of investigations that they had designed. Some of the youths actively contributed to establishing counts of benthic vertebrates—a key indicator of water quality—along the creek, with the results entering a database that the environmental group established.

Michelle's group was interested in documenting the state of the creek. For example, they used a tape recorder to record themselves describing several sites along the creek; and they took photographic images for depicting water pollution. Michelle also interviewed the mayor of the community about current policies and the possibility for changing them; and she talked to aboriginal elders about the historical role of the creek in the life of the local W̱JOȽEȽP First Nation, which belongs to the W̱SÁNEĆ group of the Coast Salish people. In the end, the participants produced an exhibit containing many photographs, exemplifying, among others, the differences between the creek where it had become a ditch draining industrial wastes and where it was in a more natural state brought about by the environmentalists' efforts in creek restoration (Figure 1.1). While standing next to the poster she had prepared with her group mates, Michelle explains to a visitor the difference her group had found between different ranges of the creek along most of its 8-km course, some of which had been turned into a ditch to drain the surrounding fields quickly and other parts that had already been revitalized:

> You would find at the creek deer, squirrels, and more animals from the forest. But in the ditch, you would find more bugs and birds, not the bigger animals, because they cannot live in those habitats. And there is no fish in the ditches; there are like little bugs and no fish. There is fish in Centennial Park, there is cutthroat trout and stickleback; and the creek is cleaner, because it is not beside the road. And people are not dumping garbage onto the creek, they dump it into the ditch, out of their cars and as they are walking by. We found much more garbage: we found pop cans, drinking things from McDonalds, French fry cases, and things like that.

Figure 1.1   Michelle talks about her research in the community regarding Hagan Creek and its pollution.

In the end, the youths' contributions to the community generally and to the database built by the environmentalists specifically came to be described in the local printed newspaper and websites.

In this sketch of a science unit, we observe how traditional organizational forms of school science have been altered in radical ways. Although the overall organization was still located in the school, the motive of participation had been deterritorialized into the community; rather than having one institutionally designated "teacher" be responsible for all youths, parents, elders, and other adult members of the community, all participated together with the youths. Participants no longer did tasks designed by a teacher but realized a piece of work that they designed themselves in response to the call an environmental activist group made for community participation. The youths were oriented to doing something for the health of the creek specifically and the watershed more generally, rather than completing however well designed tasks and tests only to put away what they have done after grades have been issued. In the course of their participation, and without any specific orientation to learn some specific science concepts, the participants were changing, growing in their orientation and practices toward the creek, in their awareness of its state and health, as active and engaged citizens in their community. As Marx and Engels articulate in the introductory quotation, the participating youths were becoming together with *what* they produced and *how* they produced it. At the same time, they were recognized for and associated with their contributions. Even though they were not held to study something specific, and even though their least concern was in constructing some identity, they were changing in and through the production of information and other things contributing to the environmental cause. In a way, their becoming was intransitive, occurring while they focused on producing something specific of value to the environment generally and something to be contributed to the open-house event more specifically.

While organizing this design experiment in the middle school, I was also in the process of conducting a long-term ethnographic effort to document science in the community. I came to realize that the theoretical and empirical grounds that a focus on teaching science to every individual might be ill conceived and that a collective view in which scientific literacy is produced cooperatively by people with *different* forms of knowledgeability contributing to problems that inherently are transdisciplinary in nature. What is important was the activism itself, not whether I had imposed upon the students remembering some related vocabulary (e.g., ecosystem, biosphere, or photosynthesis), facts (e.g., living things interact with their environment), or skills (e.g., observe and record the biotic and abiotic components in a local ecosystem) as per the official provincial curriculum for seventh grade.[2] The increasing specificity and extent of any form of knowledge makes it impossible for any human being to be an expert in multiple domains—as the proverbial Renaissance person. Instead, I realized that we needed to become knowledgeable—literally, have the *ability* to *know* at the time required rather than actually know during an exam only to forget immediately after—in cooperating on teams where no two individuals may belong to the same community of practice. I began to theorize scientific literacy "as a situated, distributed collective, emergent, indeterminate, and contingent process" (Roth, 2007, p. 377). Thus, even though Michelle did not measure stream speeds herself, did not establish frequencies of benthic invertebrates, and did not draw a graph correlating the two—aspects of the official curriculum—her participation contributed to the overall endeavor of representing Hagan Creek. It matters less what the individual does, achieves, and talks or writes about in a test. Instead, what really mattered were the changes that occurred in the community globally, transformed through Michelle's knowledgeable engagement with a really important issue: a general orientation toward the importance of stream and watershed health. From that perspective, scientists and engineers in our municipality were observed to be among the most scientifically illiterate people because they exhibited an inability to participate in processes where any concern and form of knowledge other than standard science came to bear.

AN ALTERNATIVE VIEW ON LEARNING

An alternative view on change in humans is articulated in the works of Marx and Engels. Thus in the introductory quotation, these authors suggest that what humans are—their knowing and identity—*coincides* with the content ("what") and mode ("how") of their production. That is, in producing goods, for example, while engaging in environmental activism, humans produce themselves even though they do not tend to think about their identity or their learning. Participating *is* transformation and learning.[3] By working toward a display at an open-house event that environmental activists organized in the community, Michelle did not focus on learning science content or on doing something to ("constructing") her identity. This was apparent one year later, where Michelle together with a peer and one of the environmentalists accompanying her group talked about what they remembered, which was very much organized around and associated with specific episodes and sites. In conducting interviews with aboriginal elders and politicians, documenting the health of a local creek with her camera and through journalistic description on audiotape, she also changed. But this change is not in the same direction as the one characterizing her intentions, which included, among others, interviews with the mayor and aboriginal elders. She did not make the change the motive of a conscious effort. Instead, while interviewing, Michelle changed and this change was not intended, was not the object in her consciousness. She, as each of us, is continuously becoming; and this becoming, which never ends from birth through death, is happening transversally to our intentional engagement in some project. The lines of our continuous becoming are not defined by points that connect; lines of becoming always run in a direction orthogonal to the connection between points that initially have been identified. Although such a description is quite compatible with common views on how people learn in informal (workplace) settings, it is uncommon for the views espoused by educators and educational researchers concerned with school learning. It is incompatible, as I show in the first subsection, with a constructivist theory of learning.

Constructivist Theory is a Dead End

In educational circles, learning and change are theorized in transitive ways. Thus science "students" or science are said to "construct" "knowledge," "identity," and so on. Such statements and the state of affairs these refer to grammatically are realized by a subject-verb-object structure, where the subject and object are the points connected by the verb. The verb is transitive, that is, an action passes from the subject over to an object (knowledge, identity). From an anthropological perspective on the experience of being alive, all endeavors attempting to understand learning in this way, constructivist or social constructivist, constitute a pit that psychologists and learning scientists have dug for themselves.

The traditional approach to learning is a dead end, because it completely neglects to consider that in the case of knowledge, there is a fundamental contradiction, for the new knowledge that a person is to construct is unknown and, therefore, cannot be the intentional object of a transitive verb. If the person already knew the object of construction, the new knowledge, then she would not need to construct it—in the way carpenters construct houses based on many experiences of having constructed houses before. From a pragmatic perspective on the emergence of new knowledge, the construction metaphor makes absolutely no sense. The construction metaphor also makes no sense from an anthropological view on making and the continued becoming of the maker, her materials, and her emergent object. This is so because all constructivist approaches theorize learning and identity in a transitive manner (Figure 1.2a): The subject intentionally orients toward its object; it learns (changes knowledge) or changes identity as a consequence of its intentional engagement with the object. The cause–effect model underlying this conception of living change processes has its origin in Greek metaphysics that has made it through Kantian constructivist conceptions of knowing to the present-day psychological notions of individual and social construction of science knowledge, identity, attitudes, interests, in short, of everything science.

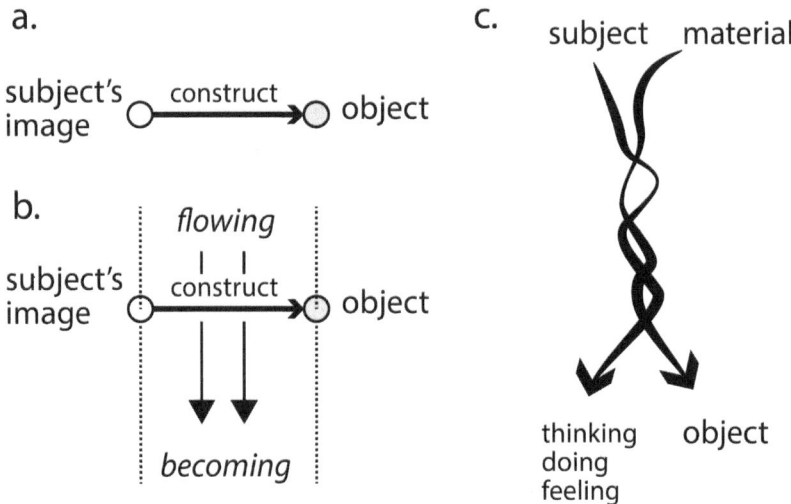

Figure 1.2   Orientations in different epistemologies. (a) The constructivist orientation is transitive. (b) The Marxist and postmodern perspectives view becoming as an intransitive process. (c) In the anthropological approach, maker and material are becoming together.

Intransitivity of Becoming

Philosophers have come to understand that the cause–effect model under-lying the transitive view of learning is a fiction, that is, the result of an a posteriori construction of causes based on the 20/20-hindsight knowledge of the effect. It has become evident in recent years that the view of human actions and their effects as the (causal) result of (mental) plans is completely inadequate to account for empirical observations. At best, the descriptive adequacy of plans of actions with respect to the situated actions that fol-low can be established *after the fact*. Continued becoming is ill described by cause–effect models because it is "not defined by points that it connects, or by points that compose it; on the contrary, it passes *between* points, it comes up through the middle" (Deleuze & Guattari, 1980, p. 359). *Becoming* therefore always is in the middle, which is not an average. It does not lie in the connection or interaction of the subject and the object (Figure 1.2a). It is not the intersection, overlap, or relation of the two: "It is the in-between, the border, or line of flight or descent running perpendicular to both" (p. 360). Such a view is represented in Figure 1.2b, which exhibits the two orienta-tions: (a) what the identified doer or maker is oriented toward, some identi-fied object, and (b) but what really matters is the continuous movement of becoming in which both subject and object are engulfed. Both emerge from this engulfment changed, each having become in the way of the other: The object has become like the person, and the person has become like the object.

When we take such a perspective on the environmental actions of the youths described earlier, what matters to understand the change and growth that the participants undergo is not the particular focus that they have chosen. The change is *not* determined by the materials included in their chosen project; and, conversely, the final product is *not* determined by the conceptions, mental structures, or prior knowledge that might be attributed to any participant beforehand. Such an orientation on the point-to-point connectors is transitive. Thinking change in terms of the intentional orientation and connection between subject and object is as if we were to concentrate on the banks of a river while losing sight of its flow. From the perspective of an anthropological endeavor interested in capturing the livingness of life while theorizing *making*, it is precisely the flow that matters, for "were it not for the flow of the river there would be no banks, and no relation between them" (Ingold, 2011, p. 14). Marx and Engels write in the introductory quotation about two forms of production, one transitive, oriented toward things used and exchanged with others, the other one intransitive, occurring along the longitudinal trajectories of becoming of materials and conscious awareness. Those processes of becoming do not start and end with particular points, do not start and finish in and with school. They are continuous, exceeding, from a purely material perspective, the life of the individual, who is only a momentary structure arising from the coming together of organic and inorganic material, itself changing along lines of becoming.

Lifelong and Life-Wide Lines of Becoming

We can push our theorizing effort even further. We may abandon the misleading orientation toward the subject and its environment involved in an *inter*action by means of which the two ends come to participate in a structural coupling. In terms of our example, we stop focusing on participants and the learning environment (curriculum and teachers) as entities that come to interact. We orient instead toward lines of becoming, the material flows that engulf people and things, flows that are material life itself. This creates heterogeneity, *métissage*, and hybridity as people and things are cobbled together in a process of bricolage (Roth, 2008). The participants, like Michelle, then are theorized as manifestations of continuous, heterogeneous biological, psychological, sociological, or cultural flows: as points on our metaphorical riverbanks that we can think as being connected to points on opposing banks. If we think about the becoming—especially when we think about becoming from an ethical perspective—then the acquisition of specific facts or the forcing of all participants to do/learn the same becomes questionable. Instead, thinking about lines of becoming allows us to consider changing flows that span the life of an individual person. We begin to theorize the role of science from a fullness of life perspective; that is, the role of science within a perspective of lifelong and life-wide becoming.

The notion of *line of becoming* is exemplified in the analogy of the wasp and orchid. Thus the "line of becoming that unites the wasp and the orchid [that] produces a common deterritorialization, of the wasp, in that it becomes a liberated piece of the orchid's reproductive system" (Deleuze & Guattari, 1980, p. 360). But this is not all. All the while the wasp is becoming part of the orchid's reproductive system, the orchid is becoming part of the wasp: It is becoming "the object of an orgasm in the wasp, itself liberated from its own reproduction" (p. 360). Traditional evolutionary thinking about the wasp and the orchid occurs in the ways in which enactivist theorists do in the educational community: Each is the transitive object in relation to the other in a process of structural coupling. In the approach proposed here, the relation between wasp and orchid is theorized transversally, in terms of two asymmetrical movements, which come together to form a unit that is becoming "down on a line of flight that engulfs selective pressures" (p. 360).

Such a view is represented in a figure that conceives of makers (science "students" or "teachers"), their materials, and their emergent objects in terms of lines of becoming intertwined for a while into a block (Figure 1.2c). While Michelle is interviewing, photographing, or recording herself describing the sorry states of Hagan Creek, that is, in the coming together (gathering) of what became her poster, there are lines of flight of materials and Michelle alike. For some time, there is a block of becoming, mutually enveloping lines of becoming, from which another Michelle will have emerged, a poster, a new form of consciousness in the community, etc. That block of becoming also includes the lines of becoming related to the many other people who were involved, including the parents driving the participants to the differ-ent sites, the director of the environmental group, the biologist who assists the youths in coming to use a colorimeter, the mayor and aboriginal elders being interviewed, and so forth. It is important to take all these other lines of becoming into account, for not only institutionally designated "students" are becoming but also the teachers (becoming more competent in teaching such units), parents (becoming more knowledgeable about the creek, educational initiatives, and learning), politicians (coming to know that environmental issues are important to participating youths), or postgraduates (becoming knowledgeable about research, learning, local environmental issues). Such becoming has to be thought in new ways. Becoming is not evolution, there is nothing like filiation, so that becoming is of a different order. "Becoming is a verb having a consistency of its own; it is irreducible to . . . 'appearing,' 'being,' 'equaling,' 'producing'" (Deleuze & Guattari, 1980, p. 292).

LEADING ACTIVITIES

The work of the Russian psychologist L. S. Vygotsky frequently is used in the Western scholarship to buttress a social constructivist conception of know-ing and learning. Close reading of a number of his texts generally not cited in the Anglo-Saxon scholarship suggests that his fundamentally Marxist ideas are (a) inconsistent with the (social) constructivist approach and (b) go well with the dynamic approach developed here. For example, he conceives of thinking and speaking as *processes* that are (one-sided) manifestations of a more encompassing process: the movement of life (Vygotskij, 1934). Thinking and speaking tend toward but never equal each other. They constitute a back and forth between the two banks of a river, the flow of which corresponds to the movement of life. In Vygotsky's approach, the object of psychology—subject, affect, intellect, thinking, speaking—is not stable but continuously changing, reaching along societal-historical lines into an infinite past and into an unknown future. What we think as the inner and outer of the person—Michelle's Self and her Others—are themselves products of these movements that Vygotsky has been considering. An important aspect of Vygotsky's theory of becoming lies in his determination of the societal relation as the locus and origin of higher psychological functions, a point that is taken directly from the works of Marx (Vygotskij, 2005). Vygotsky does not emphasize the fact that such functions arise while people (e.g., Michelle) are *in* relations with others (e.g., the mayor, elders, chaperones); instead, he specifically states that the societal relation *is* the high psychological function. That relation is not a thing, it is a continuous becoming in the back and forth of the indeterminate exchange between Michelle and those she interviews and by means of which that relation is itself defined. This theoretical view is completely incompatible with the going conception of something being constructed that afterward is internalized by the individual (like food that we share and then ingest).

Another important aspect is the distinction between *learning*, a process of incremental, continuous, quantitative change, and *development*, a crisis-like qualitative reconfiguration of consciousness. Vygotsky suggests that stepwise, developmental change arises without noticeable beginning and endpoints, and therefore *unpredictably*, while an individual is participating in some task within its reach.[4] This task constitutes a *leading activity*—leading because from this activity emerges development; it leads and comes before the qualitative change to a new form of consciousness. Consistent with Vygotsky's Marxist orientation, the important aspect of this participation is not the transitive orientation of the subject to material objects and the motive of activity but instead the intransitive becoming that occurs in the transversal direction. When some aspect of participation becomes the object of a subsequent (reflective) experience, a reconfiguration occurs in consciousness. At that instant, "the past is 'contemporaneous' with the present that it has been" (Deleuze, 1966/2004, p. 54). Thus development is the change in consciousness that occurs from Michelle's orientation while interviewing to that when she becomes aware of the different aspects of the preceding interviewing, including what has emerged from the give-and-take of the exchange relation.

The category of *experience* (*pereživanie*) actually is another example of the little-headed aspects a Vygotskian view exists in. A category denotes an irreducible unit of theory. In the case of *experience* (*pereživanie*), the unit includes the subject and its environment; and this unit has irreducible practical, intellectual, and affective characteristics. Experience (*perež*) is not a thing, not even a process with beginning and end: experience (*pereživanie*) is continuous flow. However, when an aspect stands out from this nonthematic flow, that is, when we recognize after the fact that we have had *an* experience, a qualitative change in consciousness has occurred. A previously laminar flow has gone through some turbulence and has become a different form of flow, leading to a new form of laminar flows of participation and experience is different. This orientation toward flow, laminar when things are calm, turbulent when there are crises, is captured in the notion of life as drama (Vygotskij, 2005), and in the psychologist's idea that psychology needs to be thought along the lines of drama: "psychology = drama" (p. 1030).

CODA

In this chapter, I sketch a discourse that focuses on continuous becoming that occurs in the coming together of lines of flight associated with people and material flows that come together for a while, forming a block that shapes and reshapes all lines of becoming. It does not merely replace the constructivist and other discourses—characterized by their concerns for learning and the construction of knowledge and identity. Instead, it runs in a transversal direction, concerned as it is with lifelong and life-wide intransitive *lines of becoming* that we are subject and subjected to (patients) as much as being the transitive subjects of (agents). We are not agents of our becoming or observers thereof but instead are witnesses in happenings that we only grasp to some extent after they have come to an end and therefore no longer exist.

NOTES

1 The *modern* conception of change is transitive, based on the image of the subject, who, through its action, affects the object (S→O) and, therefore, is equivalent to the cause–effect model (Nietzsche, 1954). Perspectives that have abandoned the cause–effect perspective on life are *post*modern.
2 This curriculum may be obtained from the BC Ministry of Education website: https://www.bced.gov.bc.ca/irp/pdfs/sciences/2005scik7.pdf.
3 If a traditional test does not pick up change, then this is an indicator of its lack of ecological validity, not that of a lack of change and learning.
4 So much for the Anglo-Saxon preoccupation with scope and sequence and the European emphasis on didactics.

REFERENCES

Deleuze, G. (2004). *Le Bergsonisme* (3e éd) [Bergsonism, 3rd ed.]. Paris, France: Presses Universitaires de France. (First published in 1966)

Deleuze, G., & Guattari, F. (1980). *Mille plateaux: Capitalisme et schizophrénie* [A thousand plateaus: Capital and schizophrenia]. Paris, France: Les Éditions de Minuit.

Ingold, T. (2011). *Being alive: Essays on movement, knowledge and description*. London, UK: Routledge.

Marx, K., & Engels, F. (1978). *Werke Band 3* [Works vol. 3]. Berlin, Germany: Dietz.

Roth, W.-M. (2007). Toward a dialectical notion and praxis of scientific literacy. *Journal of Curriculum Studies, 39,* 377–398.

Roth, W.-M. (2008). Bricolage, métissage, hybridity, heterogeneity, diaspora: Concepts for thinking science education in the 21st century. *Cultural Studies in Science Education, 3,* 891–916.

Vygotskij, L. S. (1934). *Myšlenie i reč': psixologičeskie issledovanija* [Thinking and speaking: psychological investigations]. Moscow, USSR: Gosudarstvennoe social'noèskonomičeskoe isdatel'stvo.

Vygotskij, L. S. (2005). *Psyxhologija razvitija čeloveka* [Psychology of human development]. Moscow, Russia: Eksmo.

2 Free-Choice Learning
What Does It Mean?

Laura W. Martin

One Friday, in a repurposed service station on a road in rural Connecticut, 25 children ages 3 to 13 were chanting: "We wish tomorrow was Mon-day. We wish tomorrow was Mon-day." They were students at a free school, part of a movement in the 1960s–1970s of grassroots alternative, cooperative, small-scale learning environments. The space was filled with things to do and divided into themed areas. The children could spend the whole day doing whatever they wanted to, including joining structured activities offered by the five teachers. The first few months, one group of children hardly came inside—they were tirelessly building forts, dams, and gullies in the woods and stream behind the gas station.

I was in the collective that developed and ran the school. It took shape in an educational study group inspired by the work of A. S. Neill, P. Freire, I. Illich, P. Goodman, J. Holt, H. Kohl, J. Kozol, and A. Graubard, among others. Members included teachers, artists, a filmmaker, a lawyer, and a city planner. The school ran for four years and served middle-class, working-class, and poor families. Many of the students were kids who were not thriving in regular schools and whose parents decided to try this experiment. Ultimately, whereas we could not claim the children who attended learned their grade-level skills perfectly, we did cure a lot of kids of their school phobias.

INTRODUCTION

The Free School movement was an effort to address the problem, particularly in poor communities, of children's emotional well-being and "the detrimental effects of authoritarian teaching techniques of public schools on qualities like intellectual curiosity" (Graubard, 1972, p. 7). It integrated pedagogy with theories of psychodynamics, human development, and moral development as well. In the spirit of the times—letting 100 flowers bloom—many interpretations of liberationist pedagogy coexisted, all with the aim of supporting children's freedom through free choice, community, and self-governance.

That was then. Now, working as an educational program designer and researcher at science museums for the past 20 years, I still live at the juncture of theory and practice. It's a tricky place to be. Science "museums" include many types of non-school learning settings, notably science centers, zoos, botanical gardens, planetariums, aquariums, natural history museums, and nature centers. They may offer exhibitions, public programs, classes, camps, outreach, teacher training, internships and more. The institutions where I have worked do not have lab meetings, classes, or study groups for staff; however, we are too busy dealing with the public from nine to five, seven days a week—or access to libraries, with the result that staff members often intuitively experiment with ideas that have already been researched, well studied, and analyzed. Connecting the more informal and freely chosen activities we offer to school agendas is often limited to providing teachers with references to state standards in order to justify field trips or to holding workshops for teachers on science methods and content.

When I have had the time to catch up on published research findings, I found academic studies that were intriguing and even inspiring, but that come from a myriad of disciplines deploying different methodological tools, different outcomes or products, and including different expert practices. Phipps (2010), for example, recently reviewed the distribution of theories underlying recent studies of free-choice learning. Among them, she observed a trend to theorize about social influences in learning while still noting a lack of agreement on definitions of learning and variation in methods tied to different theoretical frameworks. This diversity makes it hard for us to judge what is key to promoting learning and for deciding where we need to focus in our program design efforts: the subject matter? the individual? the interactions? the setting? The diversity also makes it hard to judge how we can best bridge connections to schools and opportunities for educators and further professional practices perhaps because the contexts of formal and informal learning environments are rarely compared across these frameworks.

Researchers and practitioners alike try to understand how to support science learning, arranging environments and experiences that can build on and become part of everyone's daily activities. When children were experimenting with the natural world outside our free school building, the teachers would join them from time to time and point out principles to improve their dams and rivers or features of the debris in the woods and so on. The kids sometimes paid attention and sometimes not. For me, this raised the issue of how to introduce scientific concepts that children would not or could not discover on their own, which then felt like an authoritarian imposition on them. Since then, I've realized that if nature was such a perfect teacher we would not have a need for schools and that science has both aspects to it: the creative and the received. I see the challenge now as how to design environments that effectively respect both aspects of science engagement. We do, happily, have psychological science to help us understand how creative thought and intellectual curiosity may flourish while skills and knowledge are acquired. Roth (this volume) challenges the view that science learning is "transitive" (you learn *something*), the prevailing framework in educational studies. If we look at learning as "intransitive" (you are learning), however, it leads us right to the problem of design. Can we design places, relations, and experiences that become part of who a person is?

In this chapter, I look at how the construct of "free-choice learning" is defined and how it helps us understand learning in different environments. In doing so, I address who wants to know what and what the context is of the knowledge we have about so-called free-choice learning. By framing the notion in a larger theoretical approach, I also hope to show how current thinking connects to what practitioners need from their institutions to be able to reflect on learning and apply professional strategies in their work.

First, I describe what the disciplines addressing free choice–style science learning look for, then talk about what we know in practice, research, and theory. Finally, I talk about where this leaves us. In doing so, I hope to suggest some common bases for moving forward in meaningfully supporting science learning at this point in time.

THE IDEA OF FREE CHOICE

In "Learning Outside of School," Rennie (2014) discusses what is now meant by informal learning, free choice, and informal learning settings. The earlier notion of free choice vanished with the discussion of radical school reform but has recently become associated with learning in non-school environments, that is, in settings where individuals are not following a prescribed or required sequence of study in a particular area (Lemke, Lacusay, Cole, & Michalchik, 2015), that is, "informal" educational settings.

Falk has argued that the free choice is a good term for some non-school learning because it refers to characteristics of a learning experience and is "non-consequential, self-paced, and voluntary" (Falk, 2001). By Falk's definition, free choice refers to a psychological and "relative" construct rather than a pedagogical approach. Importantly, Falk maintains it is the *perception* of choice and control by the learner that determines its consequences for learning.

"Free choice," then, is a useful term insofar as it references the psychology of why people decide to act. At the same time, it needs explicating. The use of the term may not point us deeply enough into the contexts of where people make choices and into the psychology of choosing. Ultimately, what we want to know is, Does having a choice about what activities you take up make a difference to the kind of learning that takes place? How do we make sense of the variety of questions and approaches in designing learning environments? How do we bridge school and out-of-school learning?

THE LEARNER OUTSIDE OF SCHOOL

For a long time, educators, parents, researchers, and educational institutions have been aware that out-of-school learning is consequential and of value. With the widespread advent of science centers, interactive media, after-school programs, lifelong learning programs, volunteer service projects in communities and the like, attention has been drawn to environments that people choose to participate in, often because these are seen as a potential remedy to the inequality of opportunity that persists in our mandatory, formal schooling system, particularly with respect to STEM learning.

National attention has also been focused on cultivating a nation of creative technical thinkers, with the realization that informal opportunities for pursuing hobbies and personal passions through clubs, camps, and internships should be available to learners from all sectors of society, particularly those underrepresented in science and technology fields. Government agencies in the US, for instance, are also looking at how to harness the power of choice to bridge "home" and "school" to more effectively address the goal of promoting critical and creative thinking among students. Whereas evidence does exist to demonstrate the power of informal environments for motivating and educating young people, the impact on their school performance is not as clearly understood. Because of these trends, informal learning environments are on the radar, this time with a targeted research agenda asking, do they work? Funders are looking for "evidence-based learning" of democratic and self-organized learning experiences in efforts to find programmatic solutions worth investing in, raising issues for practitioners about measuring outcomes in those environments.

WHAT IS KNOWLEDGE AND WHAT COUNTS AS KNOWING

Whereas the value and benefit of choice is acknowledged, different disciplines have approached the issue of what and where benefits accrue with different theories of what "knowledge" is, that is, with different *epistemologies*. In science learning, these include the fields of science education research, cognitive psychology, developmental psychology, social psychology (in areas of youth development, motivation, and interest development), visitor studies, project evaluation work, education policy work, sociology, anthropology, communications, learning sciences, and curriculum and program development. The myriad of approaches may confuse practitioners who have little time to develop a "big picture" sense of what works.

Some fields address the connections between schools and non-school environments, whereas others focus on a specific setting or on the subject matter of science. Each discipline addresses different and fundamental questions that touch on the concept of free choice, illuminating aspects of the importance and diversity of learning environments at this point in time.

Questions currently being asked about STEM knowledge and knowing may include the following:

- How does STEM education expand the workforce pipeline?
- How can we encourage the integration of interest-based science experiences into the regular school day?
- How do we pass along information in this society and contribute to an evolving culture?
- How do we insure equitable access to scientific thinking and 21st-century skills among all students?
- What supports cognitive development? How is understanding constructed? What mental models are being applied by learners?
- How do we design educational environments and tools, including electronic tools, outside of school repertoires?
- How does STEM learning address the crisis in Western schooling and the democratization of information in the electronic age?

WHAT WE KNOW FROM PRACTICE

We have learned a lot about free choice from practice, research, and theory. To start, we have examples of best practices. Best practices are, in a sense, local windows into learning, but they may not, by themselves, help us see how definitions of learning are evolving and how the field is constructing a culture of epistemologies. They may not explain the learning paths that individuals take in different settings and the complex histories people bring to any particular program that allows them choice. Finally, compilations of best practices do not necessarily yield insights into how we create learning relationships or how to pass along the insights from projects through professional development and sustainable planning. Best practices *are* useful, however, in helping us gauge what seems to be working in our communities at the present time.

Some of the domains that have been studied among best practices of informal activities for youth include:

- The quality of the experience (e.g., voluntary, social, mentored, with flexible outcomes; balanced power relations regarding goals and means);
- Institutional structures that support youth development (e.g., being community-based; providing leadership opportunities to youth in the programs; staff informed about youth outside programs; holding regular staff meetings; providing physical and psychological safety, supportive relationships, opportunities to belong, support for efficacy and mattering, opportunities for skill building; integration of family, school, and community efforts); and

- Factors that support STEM learning experience (e.g., hands-on learning, project-based instruction, relevance to the real world, probability of obtaining rewards from people they admire in their own communities, intrinsically interesting curricula and tasks, instructors who are near peers).

The lessons practitioners have learned over the years have a basis in research in the social sciences: anthropology, psychology, and sociology.

WHAT WE KNOW FROM RESEARCH

The roots of research on formal and informal learning can be traced to early studies comparing technical learning in Western schooling with apprenticeships, training at work, and other non-school experiences in traditional societies. Learner choice was not really the issue; the studies were about how cultures frame access to learning tools and tasks and concerned practical, often technical, activity and its consequences for individual learners. Studies of everyday thinking, though, show that critical aspects of the learning setting can be viewed from both sociocultural and individual cognitive angles.

Sociocultural Factors Affect Choice

The context of the learning matters. Differences in cognitive outcomes between formal (Western) schooling and learning outside of schools depend on: the players involved, the source of the authoritative knowledge, the applied or abstract nature of the lessons, and the goals of participation. A society's arrangements of roles and customary ways of passing along skills and knowledge through communities of practice have a large impact on how learning happens.

Learning communities are learning units. Social configurations of groups are a locus for learning so much so that a recent review of research in informal learning argues that the unit of analysis of learning in these areas be broadened "to include learning by groups and whole projects as well as by individuals" (Lemke et al., 2015, p. 3).

Meanwhile, action researchers, educators, and the public are designing alternative models of access to science through new kinds of learning communities all together, blending online environments and face-to-face environments, often in school settings but also in maker spaces, game worlds, and online citizen science collaborations.

Places of learning are workplaces, too. People learn at school, at work, at home, and in their free time in communities. Central to how free-choice learning gets carried out in different places are the circumstances under which educators find themselves. Conditions that support effective professional development for teachers are well studied, but we do not know much about conditions in informal educational workplaces where the institutional focus is on delivering programs to the client rather than on professional development.

Individual Factors Affect Choice

How learners understand the world is important for educators to understand. Researchers agree that children's thinking is constrained by a lack of information and experience rather than by a different kind of logic or reasoning than adults. Important work taking this into consideration looks at both the structure of the sources of information for children and at the changes in the knowledge base they employ/construct, often as they interact with their parents in free-choice settings such as science museums.

Motivational factors are part of learning: interest and identity. In recent years, learner motivation—including interest and identity—has entered the picture in considering how people learn and how to arrange optimal environments for learning (Bell, Lewenstein, Shouse, & Feder, 2009). When anchored in relevant community issues, science programs can support learner motivation through sustained interest and identification. Providing "authentic" science practice, engagement in out-of-school (i.e., non-judgmental and exploratory) activities, social communities, links between science and everyday life help young people make connections to school science by supporting their scientific dispositions, that is, their tendencies to act on their interests (Clegg & Kolodner, 2012).

Renninger (2010) and her colleagues define what "developing interest" in math and science looks like, demonstrating four stages that people go through when we say they show interest. This work shows the kinds of environmental support that can move someone from one stage of interest to the next so that it is tracked in context and not merely located inside an "interested" party.

A recent National Research Council report (Bell et al., 2009) considers identifying with the scientific enterprise as a legitimate outcome of science learning in informal environments. Aschbacher, Li, and Roth (2010) say "*science identity* is the sense of who students are, what they believe they are capable of, and what they want to do and become in regard to science" (p. 566) They found that family encouragement, teacher feedback and expectations, classroom experiences, and sharing with peers support positive attitudes and decisions regarding science careers. Without these supports, students lost interest and dropped out of science-related pursuits.

Identity seems to be strongly linked with social, interactive, and contextual factors, pointing to the experience of setting as important in defining who learners become. When anchored in relevant community issues, science programs can support sustained interest and identification. Gee (2000) recognizes the fact that the construct of identity is used in specific social-historical-political contexts to label people and analyze how they gain status, revealing how institutions work to shape identities through access to supportive resources. His analysis is important because he does not locate identity in a person but, rather, explicates how identity is *enacted* in our society.

WHAT WE KNOW FROM THEORY

Cultural-historical and experimental psychological insights converge as educational researchers acknowledge the effects of contingencies of reinforcement in different environments and how those determine attitude, engagement, identity, and even knowledge transfer. The former addresses the multileveled nature of human development, interaction, society, and culture; the latter addresses experimental evidence of controlling variables universally affecting human behavior. Together they explain how contexts drive goal formation and how goals influence thinking. Learning may be learning, but practice, research, and theory all conclude that the patterns of outcomes for an individual in formal and informal contexts lead to different learning later.

The effects of free choice addressed naturalistically by studies in the cultural-historical activity framework and experimentally by studies in experimental psychology are these:

Results of Practices Can Be Observed at Several Different Levels

Cultural-historical activity theory explains how development of a species, its culture, and its individuals are interrelated and can be traced on four different levels simultaneously: the organism, the individual, the society, and the species. Forces influencing learning and access to learning at the societal level are more fully researched around school settings, where studies may look at a range of ways in which teachers, schools, or school systems can introduce practices to improve student performance. Studies of systemic factors in informal settings such as science museums are notably missing.

Social Histories and Social Setting are Determining Factors

Our social and cultural experiences and our biology determine what we orient to. Our future intentions grow out of our histories, a result of what we've learned and experienced in the past. So a "choice" to do something actually represents a moment in a sequence that is not entirely observable, encompassing learners' past experiences in that context, their experience with the people and objects present in the situation, and the goals they think are the purpose of their actions. This history fuels the mental operations and attention they draw upon, which, in turn, determine learning outcomes. The outward appearance of actions, then, by two different people—like conducting an experiment or exploring an exhibit—may look the same but because the reasons for undertaking it are different, the reasoning and learning about it are not the same at all.

To get the history behind someone's thinking on specific occasions, both cultural-historical activity and operant psychological theory examine interactions or "verbal behavior": the gestures, talk, and moves that people make as they act. In the case of cultural-historical activity theory studies, inferences are made about *the form* of utterances (e.g., Are they Descriptive? Predictive? Explanatory? Referential? Questioning? Co-constructed?) and their correlation with evidence of "learning," such as the ability to solve a new problem. Experimental studies look at the *function* of the utterances according to whether they control conscious choices or result in changes of patterns in people's actions.

Goals Determine Thinking

The differences between free-choice learning and prescribed learning have to do with several features of all educational, goal-driven events. First are the origins of the goals we strive to achieve. Roth (2014), for example, describes "expansive" and "defensive" learning where the positive or negative reasons for task engagement come to characterize what exactly learners get out of an activity. Experimental psychology finds that learning in order to reduce the threat of something (e.g., bad grades) leads to a very different attitude, goal development, and internalization than does learning that is undertaken in order to generate a pleasurable outcome (e.g., a feeling of achievement).

Other experimental findings relate to whether goals are defined by others or by the learners themselves: externally versus internally defined goals have been shown to make a difference to learning in experimental work; having rules or requirements coming from others does not mean they are followed or internalized. Following rules has also been shown to result in less flexible learning. These principles are why, perhaps, in more qualitative analyses of interactions in different settings, we see that learners are highly persistent and creative when they develop their own goals for activities: They are responding to the consequences of their actions directly rather than to a teacher's instructions or "recipes."

Consequences of Actions Change Learning Patterns and Retention

"Rules" or conventions of the society or of a setting, of course, are internalized and can become internal motivators. Naturalistic studies of parents shaping children's understandings in everyday interactions and with museum visitors experimenting with hands-on devices demonstrate that verbal cues shape learners' goals and their contacts with the environment. In this way, children learn the conventions of their society and what is expected of them.

Experimental work looked at the consequences of phenomena such as shaped behavior (successively approximated through trial and error and action) versus rule-governed behavior (resulting from verbally explicit rules), also considering subjects' internal states of motivation and emotion. These studies showed that "genuine" (direct) consequences of activity are more likely to determine engagement and knowledge acquisition than rule-governed behavior.

In sum, the nature of the outcomes of the actions we take shapes the choices we continue to make as opportunities and experiences unfold. What we, the observers, call "choice" is whether an action is perceived to originate in an external source or whether it appears to derive from a source internal to the actor.

Uses of Artifacts Change Possibilities

From studies in an activity theory framework, we find that mediating artifacts—computers, exhibits, manipulatives—as well as sign systems and social groupings facilitate specific actions, but it is the cultural deployment of learning tools and the goals of their use that determine the broader activities people engage in. In several studies, concepts covered in educational books, video, and computers were internalized quite differently by schoolchildren depending on the setting in which they were encountered. If children, for example, perceive they are free to comment and react to videos viewed at home, they derive different information from the medium than they do at school.

WHERE DOES ALL THIS LEAVE US?

Free-Choice Learning Makes a Difference in Science Learning

Choice in formal, semiformal, or informal science learning environments does make a difference in the fundamental ways experience becomes part of an individual's history and shapes the goals she or he perceives or targets in different contexts. It makes a difference in the way internalization of science content occurs and in the way attitudes develop. Free choice itself never occurs outside a cultural framework, however, so we should continue to study how educators and learners participate in communities that do science, internalize the practices, and are made aware of their connections to those practices, both in school and non-school environments. Findings from experimental work on the kinds of thinking free choice supports are particularly relevant for the subject matter of science: Having choices develops internal motivators leading to flexible, creative, and extended engagement with scientific practices.

Multiple Approaches Can Be Integrated into Understanding Science Learning

Many fields are invested in understanding how people voluntarily engage in science learning and whether that energy can be harnessed for prescriptive settings. Many expectations exist for what constitutes outcomes of free-choice science learning experiences and many practical approaches exist to facilitate and support them. Is there a single science to all this? I'm not sure. There is value in continuing with a range of methodological tactics, juxtaposing the findings so they can help us work on the short-term, medium-term, and long-term components that constitute learning settings (Lemke, et al., 2015). This is a long-term, multifaceted undertaking.

Designed Science Encounters Help Bridge School and Out-of-School Learning

The thinking about free choice has shifted in my lifetime. The conversation about free schools, resisting authority in public schools through pedagogical practices, and nurturing happy, fulfilled individuals in what we now call underserved communities is largely forgotten. The broader society is worried about school performance, measuring economic return on investment, and career readiness.

There are some rays of light, however, that show promise for incorporating the positive features of free-choice learning with prescribed standards. Technology is helping to bridge learning environments, giving learners more control. Teachers are assigning online games, searching, and coding to their students, and there is a proliferation of interactive educational options for people finding their way on their own. Designed environments that encourage science encounters, such as maker spaces, citizen science clubs, and after-school enrichment programs are proliferating. These will eventually illuminate where people in our society choose to do their thinking and what they hope to become as a result of their choices. Learning trajectories may become strengthened and more individualized as A. S. Neill, Maria Montessori, and others recommended many years ago.

In making this come about, there is a missing piece to consider. If we want to leverage the cognitive and identity-conferring advantages of free-choice science learning and institutionalize them as practices, we must consider the circumstances under which educators work. Whereas teachers are becoming familiar with the concept of "connectedness" in practice, they, as well as informal program designers and educators, are not usually trained in the psychological implications of the teaching role and fundamental principles of human development in a cultural framework. The priorities for museums, community centers, after-school programs, and the like continue to be exhibits, social services, safety, and recreation. In these settings, when staff and scientists are encouraged to think about their interactions with clients and to adjust their practices to better accommodate learners, the constraints of the workplace often prevent sustainable professional development. Educators are rarely given time, space, or mentoring to reflect on the ways in which they can best encourage interest, engagement, and identity. And so the divide between research and practice continues.

CONCLUSION

We know a lot about effective environments that support learners and result in inspiration and in engagement with subject matter. Naturalistic, cultural, and experimental evidence shows that when learners control their own learning, the effects are powerful. That control depends on learners' experiences with the learning environment, their goals, and the tools available, including the human facilitators. Science learning is particularly "direct" with unpredictable outcomes and so could flourish in situations where learners control the choices they make, thereby creating powerful experiences for themselves in relation to the natural world.

We have theory about how choice works, we have research results, and, we recognize the effects in practice. Our ability to apply these insights through professional practices is determined by elements of the workplace. Formal and informal science learning institutions have yet to allow sufficient time and space for their staff members to learn about and work on free-choice environments and to understand the reasons certain pedagogical practices are effective so they can design activities to suit their particular settings. Work from the free school movement era, unfortunately, is not part of our epistemic context any more, but striving for the happiness, equity, and fulfillment for students is still relevant despite the "pipeline" and "career ready" concerns that dominate justifications for supporting new approaches to science education.

As multiple views of learning and learners proliferate, as new tools and projects evolve, bridging home, school, and informal programs, as more funding becomes available for developing these "blended" experiences, there is bound to be productive cross-fertilization. Hopefully, sharing between institutions with different educational objectives will happen in a reflective mode so that science education can develop in the richest and most engaging way for all learners.

REFERENCES

Aschbacher, P. R., Li, E., & Roth, E. J. (2010). Is science me? High school students' identities, participation and aspirations in science, engineering and medicine. *Journal of Research in Science Teaching*, 47, 564–582.

Bell, P., Lewenstein, B., Shouse, A. W., & Feder, M. A. (Eds.). (2009). *Learning science in informal environments: People, places, and pursuits.* Washington, DC: National Academies Press.

Clegg, T., & Kolodner, J. (2012). Scientizing and cooking: Helping middle-school learners develop scientific dispositions. *Science Education, 98,* 36–63.

Falk, J. H. (2001). *Free-choice science education: How we learn science outside of school.* New York, NY: Teachers College Press.

Gee, J. P. (2000). Identity as an analytic lens for research in education. *Review of Research in Education, 2,* 99–125.

Graubard, A. (1972). *Free the children: Radical reform and the free school movement.* New York, NY: Vintage Books.

Lemke, J., Lacusay, R., Cole, M., & Michalchik, V. (2015). *Documenting and assessing learning in formal and media-rich environments.* Cambridge, MA: MIT Press.

Phipps, M. (2010). Research trends and findings from a decade (1997–2007) of research on informal science education and free choice learning. *Visitor Studies, 13,* 3–22.

Rennie, L. J. (2014). Learning science outside of school. In N. G. Lederman & S. K. Abell (Eds.), *Handbook of research on science education volume II* (pp. 120–144). USA: Routledge.

Renninger, K. A. (2010). Working and cultivating the development of interest, self-efficacy, and self-regulation. In D. Priess & R. Sternberg (Eds.), *Innovations in educational psychology: Perspectives on learning, teaching and human development* (pp. 158–195). New York, NY: Springer.

Roth, W.-M. (2014). *Curriculum*-in-the-making: A post-constructivist perspective.* New York, NY: Peter Lang.

3 Attention to Content
Some Lessons From School-Oriented Education Research

Marianne Achiam & Jan Alexis Nielsen

Today, it is widely acknowledged that out-of-school institutions have the capacity to positively influence science education. Out-of-school science experiences are thought to enhance the acquisition of science at the primary, secondary, and even tertiary level; to positively affect the recruitment of students to the natural sciences; and to contribute to the creation of a scientifically literate citizenry. Despite this acknowledgment, our understanding of the precise mechanisms underlying this capacity has tended to lag behind that of the classroom or school-based research community. In this chapter, we take a step toward closing this gap by identifying the areas of out-of-school science education research where we see the strongest potential for development and by drawing on research from school-based educational research to suggest productive trajectories for such development.

THE STATE OF THE ART

The discourse of the science education research community distinguishes clearly between the science education that takes place in schools or classrooms (so-called formal education) and the science education that takes place in out-of-school settings, such as museums or science centers (so-called informal education). This distinction may have had its basis in early attempts to examine out-of-school science education using methodologies and instruments that were originally developed for classrooms. Certainly the failure of the early attempts to document science learning in the terms specified by those methodologies led the community to turn instead to reflections on what might be modalities of learning particular to out-of-school settings. The notions of constructivism and sociocultural theory had a strong impact on these reflections. As a result, influential theoretical frameworks were developed to explain how learners in out-of-school settings interact with other learners, the environment, and their prior knowledge to make meaning during their visits to these settings. The question was no longer about what an individual learns as a consequence of visiting a museum or seeing an exhibition; rather, the question became about the ways in which the environment contributes to what that individual knows, believes, feels, and is able to do (Falk & Dierking, 2000, p. 12).

As science education researchers, we wholeheartedly agree with the appropriation of constructivist and sociocultural theories to help recognize the richness and complexity of out-of-school learning environments, and to account for the learners' role in co-constructing and making sense of what goes on in these environments. However, we feel that somewhere along the line, the proverbial baby may have been thrown out with the bathwater. In other words, the "science" part of out-of-school science education may have been lost in the attempts to formulate widely applicable frameworks to understand out-of-school learning. We believe that the coarse-grained inquiries made possible by "grand theories" such as constructivism and sociocultural theory have led to a focus on content-general learning outcomes to the exclusion of learning activities that revolve around specific science content.

Although these reflections about the reasons for the current state of affairs are hypotheses, the fact remains that much contemporary research conceptualizes out-of-school science education in a way that is content general and gives absolute authority to the meanings learners may construct from their interactions. In the following, we problematize this situation, arguing that it has inadvertent but important effects on the ways in which educators, curators, and other staff members in out-of-school science education institutions think about educational design.

UNINTENDED EFFECTS

One unintended outcome of the way science education is conceptualized in out-of-school environments is the absence of specific attention to what is being learned or experienced. Consider the two contemporary conceptual frameworks shown in Figure 3.1. The generic learning outcomes or GLOs (Figure 3.1, left) are learning modalities characteristic of visits to archives, museums, and libraries identified on the basis of the multiple dimensions of learning from culture (Hooper-Greenhill, 2007). The contextual model of learning (CML) (Figure 3.1, right) is a framework to understand how learning takes place in museums as the interaction of the sociocultural, personal, and physical contexts of the learner over time (Falk & Dierking, 2000). Both frameworks have been widely influential in research in science education in out-of-school contexts, and deservedly so. Yet neither framework specifically addresses the content; that is, *what* is learned.

One might argue that neither framework was developed to guide the design of out-of-school science environments and, thus, they cannot be held accountable for design decisions made in these settings. However, due to the absence of more content-oriented models of science education in out-of-school research, and due perhaps to their compelling simplicity, frameworks such as the GLOs and the CML *are* being used to guide and justify design

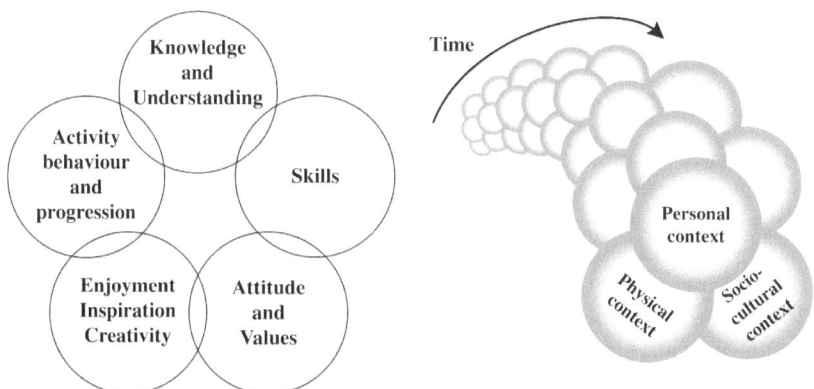

Figure 3.1 Left: The generic learning outcomes (GLOs) of visits to out-of-school cultural education institutions (Hooper-Greenhill, 2007). Right: The contextual model of learning (CML) describing learning in out-of-school settings as the interaction between personal, sociocultural, physical contexts over time (Falk & Dierking, 2000).

decisions. But perhaps more fundamentally, and irrespective of the context in which the frameworks are being used, we find it puzzling that *what* is being learned is not a part of a framework that purports to say something about learning. Indeed, research shows that models that attempt to describe learning in a general way tend to be too abstract to say anything very precise about that learning (cf. Schauble et al., 2002). Thus if we consider that science centers/museums are always *about* something (namely science), it seems a more content-specific or content-oriented approach is needed to understand and design the science center/museum experience.

Another unintended outcome of the way science education is conceptualized in out-of-school environments is what Cheryl Meszaros (2006) calls the "absolution of interpretative responsibility" on the part of the educational designers. Consider again the frameworks shown in Figure 3.1. It is easy to see how the emphasis on the learner's learning, and the near-absence of the learning environment, can cause a science center or museum staff member to misunderstand the frameworks to mean that what goes on during a museum visit is only minimally influenced by their carefully designed educational programs or exhibitions. From this point of view, it seems reasonable for those staff members to rationalize that if they have little or no impact on the visitor's learning outcomes and, indeed, if almost anything goes, how can they be held responsible for any of these learning outcomes? Meszaros goes one step further:

> By placing interpretive authority in the hands of the individual, and further, by championing the "whatever" interpretation as the final and desired outcome of the museum visit, the museum not only justifies its

> failure to communicate, but also it absolves itself of any interpretive
> responsibility for the meanings it produces and circulates in culture.
>
> (Meszaros, 2006, p. 13)

In other words, the notion inadvertently promoted—namely, that the visitors' experiences are more or less decoupled from the museum's interpretative efforts—is not only *accepted* by museum practitioners and researchers, it may even be *celebrated* by them. This can lead to a partial or complete evasion of interpretative responsibility on the part of the institution.

In sum, we have argued that although frameworks such as the generic learning outcomes and the contextual model of learning have gone a long way in orienting researchers and practitioners toward the modalities of learning that are particular to out-of-school science education environments, we believe they have two unintended effects: They may cause science centers/museums to

1. disregard the discipline-specific ways in which the scientific content is represented and experienced in their dissemination activities and
2. neglect their disciplinary interpretative responsibility toward their visitors.

WHAT ARE POSSIBLE SOLUTIONS?

To us, an important step toward solving the issues that affect research and practice is realizing that the content—in this case science—matters! Science is difficult, science is profound, and science is coming to know about the world! Most importantly, perhaps, science is what scientists do: "Scientists deploy imagination and imagery, rely upon relevant understandings, and engineer methods of inquiry suitable within particular contexts" (Ault & Dodick, 2010, p. 1101). What does this mean for the design of experiences in science centers and museums? First of all, it means that in any scientific encounter, whether it involves a scientist or a science center visitor, the individual possesses relevant understandings that can be brought to bear and has the capacity to carry out relevant trajectories of inquiry. The job of the science center/museum designer is thus to elucidate the potentially relevant understandings of their target visitors and to design encounters so that those understandings can lead to productive trajectories of inquiry. The more explicit designers are about their assumptions and intentions at all phases of the design process, the easier it is for all involved to evaluate the salience of those assumptions and intentions.

Being explicit about the intended trajectories of inquiry is maybe even more important from a classroom perspective. We do not want to overstate the similarities between classroom teaching and science center/museum experiences. Nevertheless, we are convinced that both types of experiences ought to be understood as potential *learning* experiences, rather than just experiences. At this point, we may bridge the divide between in-school and out-of-school research by drawing on the growing consensus among school-based education scholars—namely that simply presenting a topic or content is not enough to secure constructive learning experiences. Indeed, learning experiences need to be structured by a focus on *what* the learner should learn in a specific experience (Biggs & Tang, 2011). In other words, learning activities are best facilitated by a predefined set of intended learning objectives or outcomes that are explicated by the designer of these activities (the teacher in the classroom, or the exhibit designer at a science center/museum). Crudely put, the trajectories of inquiry that a visitor should go through need to have intended routes and endpoints.

To be sure, a number of scholars have argued before us that teachers or on-site facilitators ought to have clear objectives as to what visiting classes of students should learn (e.g., Anderson, Lucas, Ginns, & Dierking, 2000). But we want to make the claim that such intentions can and should be made explicit already in the process of designing exhibits. After all, the design of the exhibit is an important determinant of which potential trajectories of inquiry a visitor can go through. As such, it is not wrong to say that the designer of an exhibit can imprint her intended learning on behalf of the visitor in the exhibit itself. Thus a content-oriented approach to designing science center/museum experiences ought to not just be a matter of presenting or displaying (in a metaphoric sense) the content; but, rather, such an approach must be focused by clear learning intentions on behalf of the visitor vis-à-vis said content.

Clearly, science center/museum visits may lead to a host of *unintended* learning (e.g., as documented by Anderson et al., 2000), and of course this form of serendipitous learning should be valued. But there is no reason to fear that explicit intentions in terms of which trajectories of inquiry a visitor should go through prohibits the visitor from having learning experiences that are auxiliary to the ones intended in the design process.

Further, education scholars studying learning in formal settings are becoming increasingly aware of the massive influence that *assessment* has for learning. In this regard, it is common to distinguish between summative assessment *of* learning—which usually occurs after a teaching period—and formative assessment *for* learning—which occurs during the teaching (Black & Wiliam, 2012). Both forms seem to be key drivers, but in very different ways, for learning. Formative assessment, in particular, seems to be of paramount importance for supporting visitors in the intended trajectories of inquiry in science centers/museums. Now, in classroom teaching it will often be the teacher who provides feedback to her students about how their learning is progressing and how they might improve. But in the end, the core of formative assessment is feedback, and such feedback does not necessarily need to come from the teacher. For example, learners may to some extent give each other feedback (Asikainen, Virtanen,

Postareff, & Heino, 2014) and self-assessment also seems to be a productive strategy for learning (Boud, 2013).

In the case of science centers/museum experiences, it is clear that creative ways of giving feedback must be identified. In some respects, formative feedback may even be built into an exhibit—e.g., in the sense that the visitor's interaction with the exhibit is led through a particular trajectory by certain parameters of the exhibit. Consider, for example, the paleontology exhibit shown in Figure 3.2. Here the visitors encounter a situation where their understanding of a jigsaw puzzle is relevant for the intended trajectory of inquiry. One plausible intended outcome of visitors interacting with the exhibit could be that visitors essentially engage in the paleontological practice of piecing together the fossilized remains of extinct animals to make inferences about them. In this case the visitor's interaction with the exhibit primarily involves assembling the bones of the *Iguanodon's* foot. Visitors thus follow a trajectory of inquiry similar to that of paleontologists. As can be seen in Figure 3.2, the jigsaw-like activity of assembling the bone pieces is not a completely open-ended activity. Indeed, the outline of the *Iguanodon's* foot on the table can be analyzed as a very concrete way of providing formative feedback to the visitor: If an attempt to assemble bone pieces results in crossing the border of the outline, the visitor will (all things considered) be made aware that the assembly process is on a wrong path.

Figure 3.2 A hands-on exhibit in the Palaeontology Lab at the Royal Belgian Institute of Natural Sciences in Brussels. Casts of fossilized *Iguanodon* foot bones can be fit together using the outline of the foot on the table. Photo by M. Achiam.

Another example is given in Figure 3.3, which shows a parabola exhibit popular in many science centers. The two matching parabolas (only one is shown) allow visitors to carry out conversations over long distances, exemplifying a fundamental principle of acoustics. In this case, the visitor encounters a situation where productive trajectories of inquiry are afforded by the stool in front of the exhibit, the clear demarcation of the parabola's focus point, and the distance between the two parabolas. Although the visitor's trajectories of inquiry in this case are perhaps more distantly related to those of a real scientist than in the example given in Figure 3.2, still the parabola exhibit affords the variation of the visitor's position, distance, and sound intensity much in the same way as a physicist would carry out systematic testing and experimentation in the laboratory. In this case, too, the parameters of the exhibit afford formative feedback to the visitor. By altering her position in relation to the focal point of the parabola, the visitor will be able to get very concrete feedback to her movements in the form of shifting sound intensity.

Clearly the intended (learning) objectives of, and the ways in which formative feedback is built into, these two examples are not complex. But they illustrate quite well·how the learning of scientific content matters and the acquisition of scientific practices can be introduced as intended outcomes in science center and museum settings.

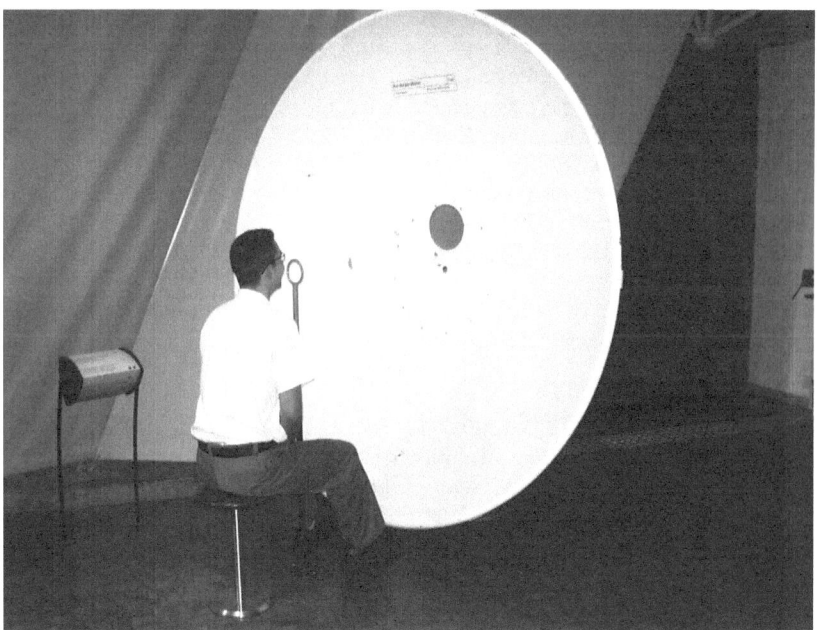

Figure 3.3 A hands-on exhibit at Experimentarium in Copenhagen. The parabolic dish focuses incoming sound waves in the focus point (the ring directly in front of the visitor) and transmits outgoing sounds (e.g., spoken words) to a twin parabolic dish that may be located many meters away. Photo courtesy of N. Quistgaard.

If at this point, the reader agrees that it is important for science center and museum *practitioners* to realize that the scientific content matters, a reasonable question would then be: What are the implications of such a realization for science center and museum *researchers*? To us, the most important implication for those of us who construct and use models for designing and understanding science experiences in out-of-school environments is the following: These models must explicitly address the scientific content in question. This entails providing the means, within our models or frameworks, for the mapping out of the intended trajectories of inquiry of visitors, and for clarifying our assumptions about the relevant prior understandings that visitors bring to the museum encounter. Indeed, the inclusion of the content as a crucial part of the museum experience would oblige researchers to explicitly address the interpretative responsibility of museums, whereas a science-specific view of the museum experience would oblige researchers to explicitly focus on how science is experienced. Thus we invite our fellow researchers and ourselves and practitioners to work toward these ends to ultimately improve the science center and museum experience.

REFERENCES

Anderson, D., Lucas, K., Ginns, I., & Dierking, L. (2000). Development of knowledge about electricity and magnetism during a visit to a science museum and related post-visit activities. *Science Education, 84,* 658–679.

Asikainen, H., Virtanen, V., Postareff, L., & Heino, P. (2014). The validity and students' experiences of peer assessment in a large introductory class of gene technology. *Studies in Educational Evaluation, 43,* 197–205.

Ault, C. R., & Dodick, J. (2010). Tracking the footprints puzzle: The problematic persistence of science-as-process in teaching the nature and culture of science. *Science Education, 94,* 1092–1122.

Biggs, J., & Tang, C. (2011). *Teaching for quality learning at university.* Berkshire: McGraw Hill.

Black, P., & Wiliam, D. (2012). Assessment for learning in the classroom. In J. Gardner (Ed.), *Assessment and learning* (2nd ed., pp. 9–25). London, UK: Sage.

Boud, D. (Ed.). (2013). *Enhancing learning through self-assessment.* New York, NY: Routledge.

Falk, J. H., & Dierking, L. D. (2000). *Learning from museums: Visitor experiences and the making of meaning.* Walnut Creek, CA: Altamira Press.

Hooper-Greenhill, E. (2007). *Museums and education: Purpose, pedagogy, performance.* Oxon: Routledge.

Meszaros, C. (2006). Now THAT is evidence: Tracking down the evil "whatever" interpretation. *Visitor Studies Today, 9,* 10–15.

Schauble, L., Gleason, M. E., Lehrer, R., Bartlett, K., Petrosino, A., Allen, A., . . . Street, J. (2002). Supporting science learning in museums. In G. Leinhardt, K. Crowley & K. Knutson (Eds.), *Learning conversations in museums* (pp. 425–452). Mahwah, NJ: Lawrence Erlbaum Associates.

4 The Museum of Pink
Retheorizing the Science Museum

Bronwyn Bevan

Several years ago, when my children were young, we were visiting the Smithsonian museums in Washington, DC, over a long weekend. After a day at the Air and Space Museum and the Natural History Museum, where my eight-year-old son buzzed around excitedly, checking in with the many rockets, planes, and other objects that he had avidly read about in books for years, my five-year-old daughter (no stranger to science or biplanes, by the way), dressed top to bottom in her favorite pink duds, turned to me angrily, hands on hips, and burst out, "I'm sick of science museums! It's not fair! I want us to go to the Museum of Pink!"

That was one of those moments for me when everything you think you know does a reverse Big Bang and collapses into a point of light. My daughter's frustrated outburst wasn't about her, it wasn't about pink, and it wasn't about the Air and Space Museum or the maybe two-dozen science museums she had visited by that time in her life. I read her remonstrance to me to be about deeper things: the role of museums, their relationship to cultural values, her formidable call for recognition and affirmation, and, ultimately, her demand for equity.

Despite its (mis)leading title, this chapter is not about gender: It is, rather, about equity and the museum and, in particular, how the intersection of formal, informal, and non-formal science education is where the deepest equity work of science museums can happen. But to understand how museums engage in equity work, there is a need to retheorize museums to broaden our view of where the work is happening and what it looks like.

EDUCATIONAL EQUITY AND THE MUSEUM

As both research and practice attending to equity in science education have expanded, the role of the science museum has become more contested. Once seen as a primary way to engage broad sectors of the population in the questions, findings, and processes of science at the cutting edge of social change, the science museum is now accused of shilling for and privileging power in its many forms. Whereas research shows that museum leaders are grappling with how to broaden their audiences (Feinstein & Meshoulam, 2014), detailed accounts of museum experiences highlight the ways in which cultural norms and practices within museums may work against inclusivity (Dawson, 2014).

The challenges are significant and reflect broader sociocultural challenges faced by all cultural institutions, including schools. But whereas analyses of who comes to the museum and their experiences there are vital, they do not tell the whole story. The visitorship lens is important and widely understood, in part, because it ties directly to the museum's economic model; but it may place too much emphasis on the agency of the museum rather than the agency of the community. Indeed, I would posit that an overreliance on structuralist rather than ecological perspectives, on the part of the research community, may be excluding some of the most cutting-edge, equity-oriented work of museums from the interpretive lens.

Equity Within and Beyond the Museum Walls

In this chapter, I draw on the work of many scholars in the cultural tradition to define equity in science education as a practice of seeing, recognizing, and leveraging learners' interests and cultural resources to support their full participation in valued activity. Pedagogically, this means making room for difference, using difference as an asset, and recognizing that difference exists in all of us.

On this view, museums seeking to support equity must deeply understand who the learners are in order to recognize interests, capacities, and commitments that can be deepened through engagement with museum resources writ large. This raises considerations with respect to who the museum "is" that is doing the seeing and the recognizing. But it is important to recognize that power and agency is not singularly located in the museum, as some structuralist perspectives would suggest. As Foucault argues, agency is fluid: It ebbs and flows through and across individuals, institutions, and communities. We need a much more dynamic and fluid way of theorizing the museum, despite the fact that it has doors and steps and can be perceived as a static thing.

As we think about how museums (i.e., the people who shape their activities) recognize and engage their publics, we need to acknowledge the ways in which they are fundamentally constrained by the limited temporal interactions most of them have with most of their visitors. These interactions last a matter of minutes to a few hours, even if they recur several times over a period of years. Moreover, people arrive and move about largely of their own volition, more or less anonymous in these public spaces. And perhaps most importantly, historically and politically manufactured social and cultural norms deeply shape who chooses to even enter museum buildings in the first place. My daughter, at age five, would have entered the Museum of Pink with exuberance. My son and me? Not so much.

But is the meaning and influence of a museum determined by who chooses to enter the building? Some would say yes, and to broaden visitor-ship they are exploring new models of engagement with their collections and spaces. For example, the Exploratorium, where I worked for 25 years, launched a program that specifically targets young adults. The After Dark events hybridize the museum's learning space to integrate ideas, concerns, and questions associated with contemporary millennial-youth culture and counterculture, for example, playful, sometimes noir-ish, engagements with phenomena such as wearable technologies, hacking, or sexuality. Lighting and sound is changed. Drinks are served. These efforts now generate the biggest attendance numbers in the month.

But while I agree that yes, we need to broaden our visitorship, of course we do, I also argue that, no, the meaning and influence of a museum is not determined by who enters the building: Without in any way diminishing the importance of objects and collections, and efforts for far more inclusive audience development, I posit that a museum is much more than a building and its contents. A physical museum is an affirmation, a recharging station, a moment of inspiration or reflection that relates to the cultural, social, and historical phenomena embodied in the museum collections that reflect, remind, and relate to the social practices in the world that led to the creation of the objects in the museum. But the museum is a moment of affirmation for people largely because of what they have brought with them to that encounter, how they have been prepared to notice, care about, engage, and find and make inspiration in the museum setting. It is outside, not inside, the museum walls that museums contribute to creating the conditions in which museum visits have meaning. How can this broader, more ecological, conceptualization of museums help us to understand how they can advance educational equity?

Ecological Perspectives

Recently, an educational improvement effort focused on *connected learning* has reinvigorated interest in ecological views on learning (Ito et al., 2013). These views help to situate and relate the role of a wide range of settings and opportunities in the learning and development of young people. Ecological perspectives on learning demand that we look beyond institutional walls or labels and focus instead on meanings and experiences that accrue over time and across settings in the lives of learners.

Ecological perspectives in education, in this sense, reference the work of U. Bronfenbrenner, who conceptualized human development as occurring within a series of nested contexts, ranging from the immediate environment in which young people interact with other people or cultural objects, such as books or games (microsystem), to the intermediate (mesosystem), the sociopolitical (exosystem), the cultural (macrosystem), and the temporal/historical (chronosystem) (e.g., Barab & Roth, 2006).

The key idea in ecological systems theory is the interpenetrating nature of the social and cultural forces that afford and shape a person's learning and development as they traverse time and settings. Deal making in government office building hallways can affect the moment-by-moment nature of how a young child is taught to read in a classroom, which may have ripple effects throughout the child's life. Network theories of ecological perspectives along with dynamic systems theory further conceptualize ecological systems as profoundly complex and dynamic.

Such views force us to reconceptualize the role of the museum (or any other agency of education, including schools): Rather than a destination point defined by its physical perimeter, a science museum is a source of energy fueling the varied interacting social systems and opportunities that make up the learning ecology. How the museum theorizes and approaches its subject matter, and the depth and breadth of its activity in the larger ecosystem, shapes broader community engagement with those ideas and phenomena, for example, through civic projects that engage public audiences, professional learning programs for classroom educators, or the use of broad digital platforms to reach learners in their homes.

Science Learning Activities and Settings

How and where do people engage with practices of science? Clearly people engage in everyday practices of science on a daily basis. Typologized as "informal" learning activities (Perulli, 2009), these everyday actions involve purposeful inquiry, data collection of a wide variety (from systematically adjusting the seasoning of food to closely observing the behavior of a sick pet), and evidence-based meaning making.

People also engage in scientific practices through many "non-formal" learning activities in which experiences are planned and, to some degree, supervised, but also allow for divergent and unexpected learner-driven directions (Lemke, Lecusay, Cole, & Michalchik, 2015). For example, a science summer camp may curate a set of experiences, but not overly determine exactly how the learner will engage with or what he or she will get from them. A school may plan a field trip with general but not overly specified goals.

Finally, science learning can happen through "formal" learning activities, at schools and also in structured classes offered in other settings. Under these conditions, learning is planned, sequenced, and scaffolded in a variety of ways (including in emergent or organic ways) toward predetermined learning goals. Success in these settings means that, whatever the path, educators intend for the learners to arrive at predetermined understandings of, or capacities to practice or advance, the discipline. Therefore, summative and even standardized evaluations can help to assess whether at a minimum the intended learning outcomes were achieved; whereas standardized assessments would be deeply problematic in non-formal or informal settings, where "endpoints" emerge in relation to productive paths that learners identify and pursue, and may vary for each individual because of the situated, learner-driven nature of the activity.

Learning *settings*, which are historically and socioculturally constructed, tend to shape or optimize different opportunities for learning (Bevan et al., 2010). In the US, after-school, youth-serving programs were founded in the late 19th century as safe havens and resources for skill building for urban, largely immigrant, communities. These principles continue to shape, in expanded youth development terms, the possibilities within today's after-school programs. Both research and practice frequently reference and typologize institutional settings as either formal (degree granting) or informal (non-degree granting). Formal or informal institutional *settings* should not be confused with formal, non-formal, or informal teaching and learning *activities*. In any given learning setting, and during any personal trajectory of learning, these different modes are constantly emerging and blending. For example, formal instruction in the school (formal) classroom is peppered with all kinds of informal exchanges that shape learning. A museum can design a particular learning program, and even quiz people at the end of it, judging success by whether or not learners noticed, remembered, and reproduced particular ideas or facts. This is formal instruction in an informal setting. Formal and informal sectors consist, respectively, of specific types of institutional settings, with institutionalized norms regarding professional practice and preparation, funding structures, organizational missions, etc. Collaborations across sectors, therefore, will integrate or blend settings and activities to produce activities that are more or less formal (with intended outcomes) or informal (with emergent, divergent, and learner-driven outcomes) depending on the goals of the collaboration.

An ecological perspective on educational equity would seek to theorize and understand how museums work across institutional sectors or boundaries. In these cross-sector efforts, museums work toward a vision of social change that can more fully democratize access and make use not only of the museum as a sociocultural resource within a community but also of the disciplinary practices and ideas that the museum embodies but that exist throughout the community (in school, workplace, home, and everyday settings). In the rest of this chapter, I share examples of what this can look like.

Realizing the Potential to Support Equity in Science Education

Museums need to make choices about how they can actively engage high leverage points in their local learning ecologies to advance an equity agenda. In this section, I share two Exploratorium examples of strategic choices and their outcomes.

The Exploratorium is a laboratory for designing learning experiences (exhibits, programs, film, books, and digital media) that aim to heighten curiosity, leverage and foster a sense of play and inquiry, and inspire people to notice, care, and ask questions about the world around them.

The museum's approach to science as a process of inquiry, and to teaching and learning as processes of inquiry leveraging one's developmental resources of curiosity and play, represents an expansive and fundamentally equity-oriented pedagogy in that it is designed to start with the observations and questions of the learner and to support them in pursuing these phenomena further. However, the equity move incumbent on the museum is to make this pedagogy accessible (both available and welcoming) and relevant to learners across all sectors, not only to those who choose to come through the museum's doors. The examples I provide are meant to highlight ways in which working across the intersection of formal and informal learning sectors—primarily through collaborative work with classroom and afterschool program educators—can help to realize the potential of museums as forces for educational equity.

Expansive Approaches to Science Across Formal and Informal Settings

Teachers are perhaps the most important influence on how different generations within a community come to understand and care about science. When science is narrowly defined and taught as a series of facts and procedures, it is likely to be of less interest to most people. When it is brought alive as a vibrant process of inquiry with social relevance and meaning, it is likely to be far more appealing and therefore inclusive. Since the museum's founding, a central impact strategy has been to work with science teachers and school administrators, across the primary and secondary grade levels, prioritizing educators from state schools serving young people from low-income communities.

The Exploratorium's programs aim to support teachers' long-term and continuing improvement of practice over a period of years. For example, secondary science teachers participate in the programs for an average of seven years, returning of their own volition to attend Saturday and summer workshops and institutes, and frequently electing to serve as mentors and coaches for teachers new to the program. The K–12 programs, providing in-depth supports to some 300 teachers a year, give teachers an opportunity to be playful learners by starting with inquiry at exhibits; moving to extended, more structured inquiry in tabletop activities that can later be used in their own classrooms; and finally reflecting on design and facilitation decisions that support or impede the channeling of curiosity and play into sustained scientific inquiry. Through these experiences, the programs seek to support science teachers' conceptual development, classroom practice, and teacher identity: They recognize and honor teachers' passion for science, their love of teaching, and their commitment to professional excellence.

Whereas the entire teacher learning effort—across primary and secondary grades, from novice to mentor to district leader—is a central equity strategy for the Exploratorium, I share here one particular strand of work.

In the late 1990s in the US, large numbers of teachers began to retire, and there were suddenly significant teacher shortages that created a wide range of emergency credential programs, meaning that a large number of new teachers with little to no formal preparation were flooding the classrooms, especially in the sciences and mathematics. Up until that time, the museum's secondary-level programs had only worked with science teachers with at least five years of classroom experience, who statistically were more likely to stay in the profession.

The sudden influx of new teachers, accompanied by alarmingly high turnover rates of up to 40–50% within two years, hit schools serving low-income communities the hardest. Veteran teachers in our programs began to ask us if we could help their novice colleagues who were struggling in the classroom. In response, in 1998, the Exploratorium created the nation's first science-specific teacher induction program. The two-year, 200-hour program bolstered teachers' mastery of science content and their classroom use of inquiry-based instruction. It also included classroom coaching and mentoring by more experienced program veterans who attended an additional 100-hour teacher leader program at the museum.

Most important, the program focused on developing teachers' identities and participation as active members of a professional community dedicated to an approach to science teaching that stressed curiosity, play, and inquiry. Thus the pedagogy of the museum and its exhibits were infused into the professional community and practice of classroom educators. Almost 20 years later, the program has worked with over 400 beginning science teachers, many of whom continue to participate in Exploratorium programs as alumni, including ultimately becoming mentors to incoming novice teachers.

A study by researchers from the University of Michigan found that teachers in the program participated in all program components at high rates, deeply valued the support they received, developed significantly stronger science content knowledge, frequently assumed key leadership roles in schools and districts, and possessed a more varied instructional repertoire compared with matched controls of beginning science teachers in other programs. A recent survey showed that, in the intervening years, 76% have taken on leadership roles in their schools, 70% report mentoring other teachers, and 61% have provided professional development for their colleagues. Most significant, the study found 91% stayed in the classroom for at least five years. Over the life of the program, 73% of graduates are still teaching in the K–12 classroom, 19% are in the field of education, 5% are not in the workforce, and only 3% no longer work in education (Heredia & Yu, 2015).

This program was designed to address systemic needs related to teacher shortages, individual teachers' needs related to lack of preparation and support, and the museum's mission to inclusively shift how science is experienced and valued within its community. The physical museum provides visitors with playful opportunities to notice and ask questions about natural phenomena. The teacher programs draw on the Exploratorium's conceptualization of science and science learning to inform and enrich science teaching and learning for generations of young people, some of whom may never step into the museum itself, or when they do they may find themselves already in conversation with the museum's approach to science and science learning. In this way, the intersection of museum pedagogy and teacher practice is a fundamental equity move for the Exploratorium.

Equity-Oriented Tinkering in Community Settings

My second example relates to STEM-Rich Tinkering in community settings. STEM-Rich Tinkering represents an approach to teaching and learning that builds on young people's curiosity and sense of play to engage them in processes of designing, testing, exploring, and problem solving their ideas. Young people come to understand and care about STEM through experiencing it as a creative and powerful means (not an end unto itself) to achieve their design goals (Petrich, Wilkinson, & Bevan, 2013).

Learning through processes of construction and making has deep historical roots in the pedagogies of J. Dewey, F. Froebel, M. Montessori, S. Papert, and others. Tinkering is a branch of making that involves creative, improvisational, playful problem solving (Resnick & Rosenbaum, 2013). Rather than step-by-step construction leading to one correct endpoint, tinkering activities by design have a range of possibilities or pathways that learners can pursue based on their varied interests, skill sets, and understanding. While engaged in tinkering, learners operate at the boundaries of their understanding. Their design goals almost always slightly exceed their technical or conceptual comfort zones; but through iterative and persistent processes of design, test, redesign, learners come to experience conceptual breakthroughs that, we have found, suffuse them with feelings of accomplishment and, therefore, a desire to continue.

Tinkering is a powerful context for equity-oriented instruction in several senses. First, it leverages the deeply human resources of curiosity and play, which all children bring to the learning activity, thus leveling the playing field in important ways. Second, it foregrounds—recognizes, honors, and builds on—the learner's ideas or understandings. It thus makes room for a wide variety of prior experiences and interests. Third, by design, tinkering activities entail multiple pathways in and through the learning activity, along with embedded opportunities for iterative designs, testing, and redesign, with no externally mandated "right or wrong." This last feature establishes the pedagogical context for helping all young people to realize success, intellectually and emotionally, as they develop, pursue, adapt, and realize their design goals.

As interest in making and tinkering has grown over the last several years, we have been pleased about the expansion of opportunities for more playful and curiosity-driven learning but also concerned that these opportunities have seemed, at first, to be provided primarily in more privileged settings, such as private schools, museums, and university campuses. Furthermore, we have found that much of the program rhetoric and research literature fails to illuminate the ways in which tinkering demands a level of intellectual and creative risk-taking, and a willingness to persist when not getting it right the first time, that entails dispositions that may have been actively discouraged over the last two decades of assessment-driven instruction in the US, particularly in schools serving low-income communities. A great benefit of tinkering is that it may foster creativity and persistence, but careful attention must be paid to inviting and supporting these dispositions, recognizing the role that educational policy has played in discouraging them.

Responding to these concerns, in 2012, we began a partnership with the Boys & Girls Clubs of San Francisco to provide tinkering programs to youth in community-based, after-school programs. Recognizing that we did not have the expertise for sustained community-based youth development, we hired a community-based educator and a postdoctoral researcher to work in partnership to implement and study the shifts and new possibilities afforded by adapting the museum tinkering pedagogy into community settings. This work was soon augmented through the addition of three new partners, the Community Science Workshops, Techbridge, and Discovery Cube Science Center, with whom we formed a research-practice partnership to study the design and implementation of tinkering in programs for youth from under-resourced communities.

Over time, the research-practice teams began to focus in particular on the kinds of facilitation strategies that supported young people's productive engagement in STEM-Rich Tinkering (Vossoughi, Escude, Kong, & Hooper, 2013). For example, the team has highlighted how to leverage the iterative nature of tinkering to support learning and to "reframe failure." The research is detailing how designing and facilitating for iteration creates opportunities for young people to troubleshoot and find success and also to complexify and deepen their thinking and ideas (Ryoo, Bulalacao, Kekelis, McLeod, & Henriquez, 2015). Iterative design also makes student thinking visible to facilitators, illuminating moments when they can support young people in their design process. It supports more equitable teaching and learning at the grain size of the learning activity.

This work with our community partners has led to the creation of a national professional development program for educators interested in implementing equity-oriented tinkering in after-school programs. In a 2015 pilot program conducted for the U.S. Department of Education and the Institute for Museum and Library Services, the Exploratorium developed a set of professional learning tools and strategies to help 25 after-school programs serving under-resourced communities introduce equity-oriented tinkering programs. Initial results have shown great success, with the sites indicating a strong desire to continue and expand their tinkering programs, along with increased demand for professional development guidance and resources to implement such programs. It is important to note that through these collaborations, the museum itself is also engaged in a process of learning how to better design and implement tinkering to advance equity both within and beyond the museum walls.

Through this work, we have been able to work with others to raise field-wide awareness of the importance of attending not only to making playful and curiosity-driven STEM learning activities more broadly available, but to making them more deeply accessible (welcoming) and relevant (responsive to the interests, experiences, and needs of the learners) to support inclusion and equity in STEM learning.

CONCLUSION

There is a long way to go before museums, and other educational and cultural institutions, are fully realizing their potential to advance equity in education. But overly structuralist approaches to theorizing what a museum is (a place) may miss opportunities to truly imagine and understand what a museum can do in its community to advance equity. Ecological perspectives widen the aperture as we examine learning developing across time and setting. It also helps us understand how different actors within a broader learning ecology can collaborate to shift possibilities (pedagogical as well as practical) for learning. It shifts our analysis away from individual actors, working behind the doors of their institutions, and instead more inclusively seeks to understand how the interactions among actors can enrich and fuel an ecology.

The examples I provided in this chapter are meant to illuminate the ways in which the intersection of the formal and informal science education sectors—in this case, in school, after-school, and museum settings—provides a powerful location for the equity and inclusion work of museums. In these community contexts, working in partnership with other educators, the museum programs help to redefine and make actionable more expansive and equitable definitions of what counts as science and what learning looks like. Developing these partnerships in response to community needs—such as an influx of unprepared teachers or the expansion of the maker movement—provides a context in which the goals of advancing alternative approaches to science and science learning are both mutualistic and relevant. Moreover, through working with educators in formal and informal settings, across multiple years, these actions have the potential to affect learning across generations of young people.

Museum visits are fleeting moments in most people's lives. They can serve as important moments of recognition and affirmation, as my daughter keenly felt when she was five and dressed in pink. But science museums that engage with their community, in sustained ways, responding to the interests and needs of their community, working within and with the community, both physically and pedagogically, learning with and from the community, can transform the way science is experienced across the learning ecology in lasting ways. Over the long term, this may change not only who comes to the museum, but who is the museum. This is the deepest equity work of a museum.

REFERENCES

Barab, S., & Roth, W.-M. (2006). Curriculum-based ecosystems: Supporting knowing from an ecological perspective. *Educational Researcher, 35*(5), 3–13.

Bevan, B., Dillon, J., Hein, G. E., MacDonald, M., Michalchik, V., Miller, D., . . . Yoon, S. (2010). *Making science matter: Collaborations between informal science education organizations and schools.* Washington, DC: Center for Advancement of Informal Science Education. Retrieved from: http://www.informalscience.org/documents/MakingScienceMatter.pdf

Dawson, E. (2014). "Not designed for us": How science museums and science centers socially exclude low-income, minority ethnic groups. *Science Education, 98,* 981–1008.

Feinstein, N. W., & Meshoulam, D. (2014). Science for what public? Addressing equity in American science museums and science centers. *Journal of Research in Science Teaching, 51,* 368–394.

Heredia, S., & Yu, J. (2015). *Exploratorium teacher institute induction program: Results and retention.* San Francisco: Exploratorium.

Ito, M., Gutiérrez, K., Livingstone, S., Penuel, W., Rhodes, J., Salen, K., . . . Sefton-Green, J. (2013). *Connected learning: An agenda for research and design.* Digital media and learning research hub. Irvine, CA.

Lemke, J., Lecusay, R., Cole, M., & Michalchik, V. (2015). *Documenting and assessing learning in informal and media-rich environments.* Cambridge, MA: MIT Press.

Perulli, E. (2009). Recognising non-formal and informal learning: An open challenge. *Quality of Higher Education, 6,* 94–115.

Petrich, M., Wilkinson, K., & Bevan, B. (2013). It looks like fun, but are they learning? In M. Honey & D. Kanter (Eds.), *Design, make, play: Growing the next generation of STEM innovators* (pp. 28–33). New York, NY: Routledge.

Resnick, M., & Rosenbaum, E. (2013). Designing for tinkerability. In M. Honey & D. Kanter (Eds.), *Design, make, play: Growing the next generation of STEM innovators* (pp. 163–181). New York, NY: Routledge.

Ryoo, J. J., Bulalacao, N., Kekelis, L., McLeod, E., & Henriquez, B. (2015). *Tinkering with "failure": Equity, learning, and the iterative design process.* Paper presented at the FabLearn Conference, Palo Alto, CA.

Vossoughi, S., Escude, M., Kong, F., & Hooper, P. (2013). *Tinkering, learning & equity in the after-school setting.* Paper presented at the FabLearn Conference, Palo Alto, CA.

5 Beyond Formal and Informal

Justin Dillon

This chapter begins with a critique of the dichotomy opposing formal and informal learning and then offers a move beyond. Possible avenues are provided where science educators may want to borrow ideas for rethinking how science learning opportunities outside of schools may contribute to learning science within schools.

INADEQUACY OF THE FORMAL-INFORMAL DICHOTOMY

In *Science Education for Responsible Citizenship* (EC, 2015), the Expert Group on Science Education "offers a 21st century vision for science for society within the broader European agenda" (p. 5). Aimed primarily at science education policy makers, the report:

> identifies the main issues involved in helping all citizens acquire the necessary knowledge of and about science to participate actively and responsibly in, with and for society, successfully throughout their lives. It provides guidance concerning increasing the participation of enterprise and industry to science education policy and activities.
>
> (p. 7)

The reports' authors use the words formal, non-formal, and informal in a number of ways: "formal, non-formal and informal settings" (p. 7); "formal, non-formal and informal spaces" (p. 21); "formal, non-formal and informal educational providers" (p. 10); "formal, non-formal and informal science education" (p. 10); "formal, non-formal and informal science education" (p. 15); and "formal, non-formal and informal science education opportunities" (p. 34). To me this is an outdated way of thinking about education, particularly the use of non-formal *and* informal. I can just about cope with the lazy shorthand that describes settings as formal or informal; but when the report talks about informal, non-formal, and informal learning, it ceases to have any academic credibility in my eyes.

I am astonished that an "expert group" uses the idea of "types of learning" as though it actually means anything.

- Formal learning—learning that occurs in an organized and structured environment (e.g., in an education or training institution or on the job) and is explicitly designated as learning (in terms of objectives, time, or resources). Formal learning is intentional from the learner's point of view. It typically leads to validation and certification.
- Non-formal learning—learning that is embedded in planned activities not always explicitly designated as learning (in terms of learning objectives, learning time, or learning support) but contains an important learning element. Non-formal learning is intentional from the learner's point of view. It can take place in museums, science camps/ clubs, etc.
- Informal learning—learning resulting from daily activities related to work, family, or leisure. It is not organized or structured in terms of objectives, time, or learning support. Informal learning is mostly unintentional from the learner's perspective.

Even a surface examination of these explanations of the three terms exposes their inadequacy. For example, "formal learning" occurs "on the job," whereas "informal learning" results from daily activities "related to work." Anyone who has spent time working with museum educators will know that many of their programs take place in an "organized and structured environment" and the learning that takes place is definitely not "unintentional." It is difficult to know how an "expert group" has allowed itself to make such basic errors in its use of terms. Indeed, the terms "formal learning" and "informal learning" are inappropriate and the term "learning in informal settings" has some utility:

> The use of the terms "formal" and "informal" does not mean that the processes of learning that take place in out-of-school settings are any different to those that occur in school: rather the terms simply refer to the site of learning. Indeed, the nature of learning—that is, the relatively permanent change in thought of behaviour that results from experience—is the same wherever it takes place. Moreover, thinking of learning as being defined by the location in which new experiences are presented is contrary to our understanding of learning as being cumulative and, as such, straddling place and time. Nonetheless, it is important to note that the learning that occurs, or is at least initiated, in informal settings is affected by a number of influences distinct from those that usually operate in schools. Such influences include the impact of novelty, for example, with respect to a visit to an imposing museum; the lack of any externally imposed curricula and thus the freedom to follow

one's own route of choice; the particular nature of the facilitation; and the social context in which the experience occurs.

(King & Dillon, 2012, p. 1905)

If we forget the imprecise and confused use of terms in the EC report and turn to what the report is recommending, a pattern emerges:

Collaboration between formal, non-formal and informal educational providers, enterprise and civil society should be enhanced to ensure relevant and meaningful engagement of all societal actors with science and increase uptake of science studies and science-based careers to improve employability and competitiveness.

(EC, 2015, p. 10)

Transforming new ideas gained from research into useful knowledge, products and services depends upon deeper and on-going connectivity between schools and non-formal and informal learning environments, families, enterprise, civil society and government.

(p. 18)

Collaboration between science educationalists, formal, non-formal and informal education providers, research centres, enterprise and industry and other professionals can play a vital role in increasing interest in science and science-careers.

(p. 23)

The "21st century vision for science for society within the broader European agenda" is one that sees greater collaboration between schools, museums, or science centers and business/enterprise/industry. The end product will be more people choosing science courses and therefore (it is argued) a more competitive Europe able to make and sell "knowledge, products, and services." The sort of actions that might take place are identified and include the development of

Innovation Hubs that link formal and informal science education with business and enterprise, SMEs, and civil society organizations at the municipal and regional level in order to:

- Foster, share, and apply science and technology research to different genres of enterprises, e.g., start-ups, SMEs, corporations;
- Encourage mentoring across different groups in order to take full advantage of science and technology;
- Facilitate mainstreaming of innovations from key enabling technologies (KET) and fields, such as health education, climate change, and environmental education.

(p. 32)

To some this is the neoliberal agenda writ large with a token nod toward social justice and transformative education. Even when the "expert group" talks about innovation in science education, it is clear that a wider group of stakeholders than normal would be involved in setting the agenda:

> Support the co-creation of innovative curricula, with defined learning outcomes involving teachers, teacher educators, researchers and representatives from enterprise and civil society.

> (p. 32)

What impact the report will have on policy makers is hard to predict. Collaborations between science centers and schools are nothing new—the *Science Education in Europe: National Policies, Practices and Research* report noted that

> School partnerships, science centres and similar institutions all contribute to teachers' informal learning and may provide valuable advice. Science centres in several countries also deliver formal CPD activities for teachers.
> (Forsthuber et al., 2012, p. 10)

Already in 2008, I noted that the experience of science education should be offered in "formal and informal contexts for learning. A single encounter with a science-based activity post-14 is unlikely to have a significant impact. What is required is a continuum of educational experiences of science from an early age."

MOVING ON FROM "FORMAL" AND "INFORMAL"

Problems with the terms "formal" and "informal" have been noted for a number of years. An early look at what the growing numbers of hands-on interactive science centers offered learners included a table of the characteristics of formal and informal learning (Wellington, 1990). However, such a dichotomy was asking for trouble and it was dismissed as "overly simplistic" (Hofstein & Rosenfeld, 1996, p. 89). More insightful, perhaps, is the attempt to frame "informal learning" in terms of two dimensions: whether "encounters" are accidental or deliberate, and whether the learning that resulted was intentional or unintentional (Lucas, 1983). It is arguable that any learning that happens in an informal setting is influenced by a number of factors beyond those usually found in schools:

> Such influences include the impact of novelty, for example, with respect to a visit to an imposing museum; the lack of any externally imposed curricula and thus the freedom to follow one's own route of choice; the particular nature of the facilitation; and the social context in which the experience occurs.
> (King & Dillon, 2012, p. 1997)

A second difference between learning in schools and learning in other settings is that, as a result of short- or long-term interest, learners are motivated to engage with ideas and phenomena. Being able to select and subsequently engage as a result of personal interest fosters intrinsic motivation that in turn can lead to deeper learning (Csikszentmihalyi & Hermanson, 1995). As time has gone by, I have used the terms formal and informal less and less. If context is important in learning, and there does not seem to be much doubt about that, then we need to be specific about what the context is and describing something as "formal" or "informal," as I demonstrated earlier, does not really help anybody.

TOWARD A BLENDED PEDAGOGY

The affordances of science centers and museums to expand the opportunities to learn about science have been documented for decades. What has become increasingly interesting is to find out how learning can be enhanced by engaging with ideas and objects outside the science classroom—online, in books, on the TV, or in other educational sites such as museums. A number of projects have looked at ways in which museums, science centers, and schools can work together to provide a new model of science education. I am also interested, though, in the extent to which pedagogical strategies can be more widely distributed across a number of contexts. That is, what can teachers of science in schools learn from the pedagogical approaches of museum designers and educators and vice versa? This border crossing I would see as moving toward a blended pedagogical approach.

It has been reported that the skill set of school science teachers varied from that of museum science educators:

> Unlike in formal settings, where teachers come to know their students over the course of a year, educators in informal settings must attempt to gauge learners' interests and abilities and modify their interactions accordingly in the space of just a few minutes. To this end, informal educators require a set of skills that are quite distinct from those of teachers (Tran and King, 2007). These skills also include the use of specimens, objects and exhibits as instructional media and the use of particular modes of talk—re-voicing, repeating, summarising—to guide, structure and scaffold learner engagement.
>
> (King & Dillon, 2012, p. 1906)

However, I now wonder about the extent to which school science teachers might adopt these skills. King's PhD thesis actually provided a pedagogical language that begins to make this project more feasible: re-voicing, repeating, and summarizing. Unlike ballet and the martial arts, teaching lacks a vocabulary of pedagogy. Whereas ballet has the *plié* and the *battement* (both of which can be further refined), science education labors with its practical work, its demonstration, and its working with small groups. This lack of a pedagogical vocabulary makes it difficult to articulate what is "museum pedagogy" beyond the rather generic use of "specimens, objects, and exhibits."

Education, of course, is not just about pedagogy. Schools are schools and museums are museums. They have different purposes and different stakeholders; they differ in governance structures and funding. However, the parallels between them are increasingly obvious. For example, educators in both might share a view that ascertaining what learners know is a pretty useful step in teaching them something new. This approach is common in schools and increasingly so in museums. Museums need to be supported in choosing their own learning paths to enable personal meaning making (Hein, 1998). Some guidance is essential—in museums, that comes from the design of exhibitions and programs from the explainers. In schools, the support might come from the teacher, the textbooks, or from other learning media. In both schools and museums, effective education is likely to challenge leaners' beliefs if they have acquired some of the common misconceptions/alternative conceptions (depending on your viewpoint) that are all too common.

Museums have long been aware of a number of other factors that affect how visitors learn and engage. Learning is affected by the interaction of their personal context with their sociocultural and physical contexts. Popular with museums is the *contextual model of learning* framework, which highlights the importance of a visitor's motivation and expectations; their prior knowledge; and their experiences, interests, and personal choices in the museum (Falk & Dierking, 2000). Sociocultural views of learning applied to museums take account of the role of accompanying family members, the presence of museum explainers, and the design of exhibitions (among other factors). Many schools would also consider such a sociocultural view of learning to be central to how they go about their business.

PEDAGOGICAL BORROWING

Chelsea Physic Garden

In terms of borrowing pedagogies, at the simplest level, I offer an idea that incorporates simplicity with relevance. On a visit to Chelsea Physic Garden in London several years ago, I came across an innovative way to help students to see where food comes from and, in particular, the value of plants (Figure 5.1). The head of education, Michael Holland, had been getting visiting elementary school children to grow plants in food packaging so that, for example, apple seeds were planted in an apple juice container. This simple yet powerful idea is one that any school could take up themselves. At the other end of the spectrum are the schools that have integrated "museum learning" into their educational philosophy and practice.

Figure 5.1   Plants growing in related food containers, Chelsea Physic Garden.

Langley Academy: "Where Every Day Is Like a School Trip"

Over the past few years, I have been working with the Langley Academy—a high school not far from London's Heathrow Airport. Langley specializes in science and has five additional foci of museums, internationalism, sustainability, cricket, and rowing. The Academy was inspired by the New York City Museum School founded in 1994. The Museum School is significantly smaller than Langley (which has over 1,000 students), but it had a more regimented curriculum arrangement in that its children spent at least two afternoons a week in museum galleries.

Langley was set up in 2008—it took over from a school that had been closed down following a visit from the national school inspectorate. The focus on museum learning comes from the influence of the nearby River and Rowing Museum. The aims of the Academy are reasonably standard apart from its commitment to "museum learning":

> Through our innovative approach to museum learning, we aim for our students to become:
>
> - successful learners
> - inspired and culturally aware
> - secure in their sense of community and identity.

In terms of what constitutes museum learning, the Academy's website explains their objectives as follows:

> Museums are gateways to real things, real stories and real people; museum collections make learning meaningful for students. With museum learning in the DNA of the school we will support our aims by informing how: students have a voice
>
> - faculties work together
> - we teach and learn
> - we use our spaces
> - we make community links
> - we celebrate our achievement[1]

The cost of setting up the Academy was substantially above the national average and it benefits from an innovative design. The atrium floor (Figure 5.2) is specially strengthened to take the weight of larger exhibits, and there are museum-quality display cases around the building. The aim is to integrate museum learning across the Academy. As the Academy's first principal, Annie Renouf Donaldson, explained:

> Our museum programme is not a bolt-on, it's at the heart of what we do. Some school trips to museums are just a one-day treat, a nice outing at the end of the term. Our own museum, our work with real museum objects in our classes, our visits to museums, these are stitched into the fabric of what we do in every class, every day.
>
> (Guardian, January 5, 2010)

The museum learning approach is integrated into the Academy in various ways. The Academy's vision could be that of a museum: curiosity, exploration, and discovery. Teachers may use a lesson-planning model to support inquiry-based learning techniques and encourage the use of museum objects, many of which are borrowed from museums both local and national.

The students' role in their education is given special attention, and there is a museum council that consists of a group of students who support museum learning activities. This group meets weekly and selects objects and themes for exhibitions and advises on new approaches. Another mechanism for involving students is that there is one student who sits on the school council with special responsibility for museum learning.

In the past, teachers have been provided with in-service training on the theory and practice of museum learning. The training involved case studies from both schoolteachers and from the museum sector, and the Academy has substantial links with other local schools and museums.

In terms of the Academy's governance, one of the governing bodies with substantial museum experience is nominated as the link governor. The Academy's

Figure 5.2   The atrium of the Langley Academy.

director, responsible for teaching and learning, has the high-level responsibility for museum learning. Finally, there is a museum learning team that consists of the head of museum learning and a museum learning officer.

Just to give a sense of what the Academy actually does, between February and May 2014, the Academy carried out the following events and activities:

- worked with 19 museums, archives, collections, and education services/organizations, ran 21 visits as part of museum learning projects;
- supported 127 on-site museum learning activities via object handling in lessons, a science week, curriculum-linked activities, the museum club, the museum council, and an enrichment activity (a debate club);
- launched the Arts Council strategic funded "Stronger Together" project;
- hosted a successful (annual) museum learning conference attended by 135 people;
- built relationships with three new organizations: National Army Museum, The Story Museum (Oxford), Oxford Castle; and,
- created 624 places for students on trips and 2,946 places for students in on-site activities, totaling 3,570 individual museum learning "experiences" for the Academy.

Whereas the Langley Academy provides an example of what can be done, the question is, to what extent can it be replicated? To answer the question, the Academy gained funding for its "Stronger Together" project that supported the development of collaborative projects between a group of teachers from Langley and nearby schools and museum professionals from different organizations. The project ran for one year (April 2014–2015) and 11 innovative partnerships were created involving 32 teachers and museum professionals and several hundreds of students.

In terms of the lessons that were learned about the potential to replicate the Langley experience, it emerged that such a partnership model requires individuals who are highly motivated and willing to engage with this type of novel school/museum practice, who are capable of establishing good professional relations, and who can capitalize on each other's expertise in a productive way. External mentorship played an important role in mediating partners' relations and it was not clear whether if this type of partnership would be possible without having that external support.

The partnerships were time consuming and involved a substantial amount of administrative work. They also required fairly regular communication among partners. One issue was that the teachers and the museum professionals had different ways of working and inhabited different institutional cultures. It also seemed to be the case that a threshold level of organizational support was a condition for successful partnerships. Teachers and museum professionals needed their managers to give them the freedom to engage in partnership with other institutions.

More obvious issues of staff training and funding also influenced the success of the partnerships. Overall, though, my sense is that there is enough evidence that what has been set up at Langley can, to some extent, be replicated elsewhere. The barriers to implementing some degree of museum learning in schools in England, at least, are not insurmountable.

CONCLUSIONS

I begin this chapter by critiquing a recent European Union report that seemed to offer a simplistic and erroneous model of science learning. The expert working group that put the report together seems to have been constrained in its thinking by policy imperatives from the European Commission. One would hope that in their heart of hearts that certainly the science educators among the group would see that there is more to a 21st-century vision for science for society than greater collaboration between schools, museums, or science centers and business/enterprise/industry leading to more people choosing science courses and a more competitive Europe able to make and sell "knowledge, products and services," which is a rather neoliberal and instrumental view of the purpose of education.

My experience of working with schools, museums, science centers, and botanic gardens over many years, and often funded by the European Union, is that, at its best, collaboration may lead us toward a blended pedagogy, whereby schoolteachers are able to learn from the best that museum education has to offer. Simply put, museum educators have to grab the attention of their visitors who they do not know nor may ever see again. They do that by having extraordinary objects that children may never have seen before or, more often, ordinary objects that they can make compelling through the stories that they tell.

I do not see that the descriptors formal and informal have much utility in the new science education that I envisage. To me it is important that students experience as much as they can during their education whether that be effective teaching in schools, visits to museums, science centers, botanic gardens, and farms, as well as art galleries, concert halls, and the country-side. Society has a moral imperative to provide all young people with diverse and enriching memories, as well as understanding of how the physical and natural world works.

In closing, I am reminded of the words of the director of the London Natural History Museum, which recently announced that it was going to replace Dippy the Dinosaur, beloved by children since it was put in the Museum's Central Hall in 1979. The museum has decided that it will replace the plaster cast of a fossilized dinosaur with the authentic skeleton of a blue whale—a species that was almost driven to extinction by human activity. The director, Sir Michael Dixon said:

> As the largest known animal to have ever lived on Earth, the story of the blue whale reminds us of the scale of our responsibility to the planet. This makes it the perfect choice of specimen to welcome and capture the imagination of our visitors, as well as marking a major transformation of the Museum. This is an important and necessary change. As guardians of one of the world's greatest scientific resources, our purpose is to challenge the way people think about the natural world, and that goal has never been more urgent.
>
> (Daily Telegraph, 2015)

The common purpose of school education and museum education could not have been articulated better. The only way forward, as I see it, is toward a shared vision between school educators and museum educators and a commitment to understanding and then blending the pedagogical approaches of schoolteachers, museum program developers, exhibition designers, and explainers.

NOTE

1 See http://www.langleyacademy.org/.

REFERENCES

Csikszentmihalyi, M. & Hermanson, K. (1995). Intrinsic motivation and museums: why does one want to learn? In J. Falk and L. Dierking (Eds.). *Public institutions for personal learning* (pp. 67–77). Washington, DC: American Association of Museums.

Daily Telegraph. (2015). *Natural History Museum dinosaur "Dippy" switched for blue whale skeleton.* Retrieved from: http://www.telegraph.co.uk/news/science/science-news/11375088/Natural-History-Museum-dinosaur-Dippy-switched-for-blue-whale-skeleton.html (accessed on September 20, 2015).

Européan Commission (2015). *Science education for responsible citizenship.* Report to the European Commission of the Expert Group on Science Education. Brussels: Directorate-General for Research and Innovation Science with and for Society.

Falk, J. H., & Dierking, L. D. (2000). *Learning from museums: Visitors experiences and the making of meaning.* New York, NY: Altamira Press.

Forsthuber, B., Motiejunaite, A., & de Almeida Coutinho, A. S. (2012). *Science education in Europe: National policies, practices and research.* Brussels: Education, Audiovisual and Culture Executive Agency, European Commission.

Guardian. (2010). *Langley academy, where every day is like a school trip.* Retrieved from: http://www.theguardian.com/education/2010/jan/05/museum-school-langley-academy (accessed September 20, 2015).

Hein, G. E. (1998). *Learning in the museum.* London, UK: Routledge.

Hofstein, A., & Rosenfeld, S. (1996). Bridging the gap between formal and informal science learning. *Studies in Science Education, 28,* 87–112.

King, H., & Dillon, J. (2012). Learning in informal settings. In N. M. Seel (Ed.), *Encyclopedia of the sciences of learning* (pp. 1905–1908). New York, NY: Springer.

The Langley Academy. Newsletter, Issue 1, October 2015. UK: London. Retrieved from: http://www.langleyacademy.org/ (accessed December 18, 2015).

Lucas, A. M. (1983). Scientific literacy and informal learning. *Studies in Science Education, 10,* 1–36.

Tran, L. U., & King, H. (2007). The professionalization of museum educators: The case in science museums. *Museum Management & Curatorship, 22*(2), 129–147.

Wellington, J. (1990). Formal and informal learning in science: The role of the interactive science centres. *Physics Education, 25,* 247–252.

Part II

Learning Science in Diverse Settings

The science education literature is dominated by research in classroom settings. However, a quick search in the Thomson Reuters Web of Science with the search term "informal science" reveals that there has been some interest in the science education community in learning that occurs in out-of-school settings for over 30 years (e.g., Maarschalk, 1986). Initially there is a slow uptake of this area of research. The 1997 special issue of *Science Education* (Vol. 81, No. 6), which focused on informal science, may have constituted a turning point. At that point, as the guest editors point out, the dialogue involving school science educators, researchers, and educators in informal settings was only in its beginning. The introduction to that special issue defines "informal learning" as referring "to activities that are nonsequential, self-pacing, nonassessed, and often involving groups" (Dierking & Martin, 1997, p. 629). Many of the studies in this special issue investigate museum learning. In the decade that follows, an increasing numbers of studies in this new area can be observed. That is, researchers have increasingly exhibited interest in the science learning that takes place in science centers, natural history museums, aquariums, botanical gardens, zoos, scientists' laboratories, public settings, the family environment, science festivals, pop-up science cafés, and other everyday life settings.

It is well documented in the recent literature that these settings offer unique educational environments and provide exciting opportunities for science learning. Dating back in part to the 1970–1980s, there exist many studies that focus on cognitive aspects of learning with the use of quantitative research methods in the context of such environments. In the past decade, there has been a shift to the use of sociocultural and qualitative approaches to research that examines the social and affective domains of science learning within a diverse set of out-of-school settings. The findings of these studies provide compelling evidence about how young people engage with and learn science in their everyday lives in a myriad of ways and kinds of social interactions. The chapters of this section offer concrete examples of how people engage (or not) with science in a variety of places and how different kinds of approaches, discourses, curricula and resources, tools and devices mediate or hinder (see, for example, chapter 7) their experiences with science.

Irene Rahm explores in "Videomaking Projects in STEM After-School Clubs: A Space-Time Analysis of One Production and Its Implicit Dialogue Among Identities, Ways of Knowing and Doing in and with Science" (chapter 6) the creative productions of video documentaries of science inside an after-school science club that entailed a partnership between the University of Montreal, schools, and community organizations. The chapter begins with a description of the production of a collaborative video among nine youths in a club that met weekly at lunchtime over a 12-week period. In doing so, the author documents vertical learning and traces connections these youths made over time among moments of meaning making in the club that year. She also describes horizontal learning and the cross-setting navigations university students engaged in with the youths that came to constitute meaning making of science and the video production process in yet other ways. By invoking a space-time lens, as the author argues, she moves beyond the after-school club "container" discourse of learning to show how learners make connections across learning environments and how opportunities to learn are organized and accomplished through travels across multiple places. In doing so, as the author discusses, the chapter offers rich insights into how learning unfolds and how learners are positioned and simultaneously author a sense of self in science and the world through engagement with science locally and globally.

In chapter 7, "When Science Is Someone Else's World," Emily Dawson draws on data from what became an ethnographic study carried out with four grassroots community groups (i.e., Sierra Leonean, Somali, Latin American, Asian) attending group meetings, such as dances, festivals, picnics, and informal science education visits in London to explore how they engaged (or not) with science in the course of their lives. The author briefly describes attitudes and experiences of science across all four groups, followed by a more detailed account from one person, Fatima, a member of the Somali community group. Fatima's stories and experiences provide evidence of her exclusion from informal science education. As the author argues, we must develop forms of research and practice that seek to disrupt rather than reproduce patterns of privilege and disadvantage in relation to science.

In chapter 8, entitled "Teachers in the Outdoors: Bridging Formal and Informal Practices," Tali Tal, Mordechai Aviam, Rachel Levin Peled, and Nirit Lavie Alon offer a concrete example of an outdoor inquiry-based program offered to teachers in Israel. The authors describe three outdoor inquiry activities that were the core focus of the program. The activities were in the fields of ecology (i.e., the study of the natural world and relationships with human impacts), sociology (i.e., the study of human societies), and archeology (i.e., the study of ancient cultures' material world). In this chapter, the authors offer detailed descriptions of the three activities;

they discuss the difficulty in attempting to make a distinction between formal and informal aspects of education; and they also provide evidence of how teachers' views of outdoor learning had developed because of their participation in the program.

Shifting contexts, Phyllis Katz presents in "The Long-Term Influence on Three Classroom Teachers of Leading an After-School Science Enrichment Program: When Identities Converge" (chapter 9) the cases of three classroom female teachers who chose to add periods of after-school science enrichment teaching to their work. The context for this study was a voluntary after-school science enrichment program in the US, where parents, graduate students, scientist volunteers, and others who did not primarily identify as science teachers participated. Through data collected via interviews, the author shows that for these women, expanding their participation in the science teaching community of practice had benefits for them as classroom teachers and suggests a continuum throughout their lives and their identities as they learned and taught about how the world works.

Chapter 10, "Integrating Mobile Technologies into Outdoor Education to Mediate Learners' Engagement With Nature," presents recommendations from research on how to integrate mobile computers into outdoor learning opportunities. In this chapter, Lucy McClain and Heather Zimmerman show how technologies can support learners to observe and learn about local plants and animals through a case study of one family's engagement patterns with technology and nature. Drawing upon data from this case study, the authors exhibit the support that four design guidelines for mobile computer integration into outdoor informal programs provide for science learning and engagement: (a) place-based observational questions, (b) place-based textual prompts for focusing observations, (c) drawing activities to record observations, and (d) place-based images used to identify biota in the outdoors.

In chapter 11 ("The Desired Role of Scientists in Bringing Authentic Scientific Practices into Classrooms: From Post–Cold War Reforms to Next-Generation Scientists"), Asli Sezen-Barrie discusses the long-term impacts of the scientist-centered and top-down curriculum developed following the cold war era. In so doing, she offers examples of recent projects to discuss the merits of scientist-school partnerships on creating effective learning environments under four headings: (a) improving learning sciences research, (b) responding to challenges in quality of teachers' scientific knowledge, (c) creating opportunities for meaningful science and authentic science experiences, and (d) creating equitable teaching environments.

As evident in this brief overview, the chapters provide an array of perspectives on learning science across very different settings expanding the initial 1990s' focus on science museums; and they provide evidence of a variety of programs, approaches, and tools that apparently foster learners' experiences in and with science. In so doing, the authors offer fresh perspectives on how people learn science in school and out-of-school situations; and they also discuss the challenges that some people face in engaging with science. As such, the chapters provide a significant contribution to existing research on how, where, and why (or why not) people learn science.

REFERENCES

Dierking, L. D., & Martin, L. M. W. (1997). Guest editorial: Introduction. *Science Education, 81,* 629–631.

Maarschalk, J. (1986). Scientific literacy through informal science teaching. *European Journal of Science Education, 8,* 353–360.

6 Videomaking Projects in STEM After-School Clubs

A Space-Time Analysis of One Production and Its Implicit Dialogue Among Identities, Ways of Knowing and Doing in and With Science

Jrene Rahm

> Any human act that gives rise to something new is referred to as a creative act, regardless of whether what is created is a physical object or some mental or emotional construct that lives within the person who created it and is known only to him.
>
> (Vygotsky, 2004, p. 7)

Many youth today are engaged in participatory cultures and affiliated with online communities, engaged in the production of creative forms of media that are shared or pursued collaboratively and posted through blogs and podcasts. Creative engagement with media is essentially about youth voice. Through the creative remixing and blending of different cultural artifacts that are picked up, transformed, and shared in new form, youth express their own understandings of their world and themselves or voice concerns for actions that are meaningful to them and their communities. The positioning of youth as creative knowledge producers and authors of their world grounded in youth voice has a long history in progressive visions of education and is also captured well by Vygotsky's (2004) notion of a creative act, which he distinguished from the reproductive act.

In this chapter, I build on that literature and explore the creative productions of video documentaries of science inside an after-school science club that was designed according to the Fifth Dimension Model (Cole, 2006). It entailed a partnership between the university, schools, and community organizations. Students from the university animated and guided youths' video productions after school in a club that was driven by youths' interests in science. The students' experiences with youth in the club were then noted in a journal and discussed in a course at the university that was part of their teacher education program. It led to an interesting dialogue around theory and practice. Through partnerships with community institutions, we could also offer field trips and a summer internship to youth from the club.

The clubs were part of an ever-increasing and vast network of after-school science practices at the elementary and high school level in the greater Montreal area. What set our club initiative apart was its focus on inquiry science and student interest–driven practice achieved through project-based science and video production.

I begin the chapter with a description of the production of a collaborative video among nine youth in a club that met weekly at lunchtime over a 12-week period. I document vertical learning and trace connections youth make over time among moments of meaning making in the club that year. I also describe horizontal learning and the cross-setting navigations I engaged in with youth that came to constitute meaning making of science and the video production process in yet other ways. By invoking a space-time lens, I move beyond the after-school club "container" discourse of learning and show how learners make connections across learning environments. I show how opportunities to learn are organized and accomplished through travels among multiple places. In doing so, the chapter offers rich insights into how learning unfolds and how learners are positioned and simultaneously author a sense of self in science and the world through engagement with science locally and globally.

I also focus on the kinds of relationships learners developed over time with the different learning environments that they came into contact with through the video production project in the club. I was guided by Vygotsky's notion of *perezivanie* and its translation and meaning as an experience colored by affect and the unity between the characteristics of the learner and the environment. Both continuously change and are reworked, making evident the need for a holistic vision of learning and identity grounded in the unity of its affective and intellectual processes. Hence I was interested in the ways the video production and its implied spatial travel opened up new opportunities for dynamic relationship building among individuals and environments, resulting in new ways of thinking about science, norms, and values tied to scientific work and affect in science (appreciation of beauty of nature, development of feelings attached to nature or animals, etc.). As noted by Vygotsky years ago, "thought has its origins in the motivating sphere of consciousness, a sphere that includes our inclinations and needs, our interests and impulses, and our affect and emotion" (Vygotsky, 1987, p. 282). Such a vision was at the heart of the design and study of the science club. I wanted to understand how that holistic vision of thought and emotion was worked out in situ, took shape, and was contested across space over time and in turn constitutive of youths' relationship with and understanding of science. It led to the following two research questions that I address in this chapter:

1. How are opportunities for learning and identity work in science organized spatially and over time in the context of one video production project in the club?

2. What kinds of relationships do the learners develop with the different learning environments they encounter through the video production process?

THE SCIENCE-TECHNO CLUB: TRACING THE PRODUCTION OF ONE VIDEO PROJECT

I draw on qualitative data collected in the context of an action research project that led to the development of science clubs in two different high schools in underserved communities with youth that I ran for three consecutive years (2011–2014; field notes by researchers and instructors, video ethnography, artifacts, interviews of youth and instructors). The design of the science clubs was inspired by the work of Furman and Calabrese Barton (2006), who experimented with science video documentary making in an after-school setting given their commitment to creating a space for youth voice–driven forms of engagement with science. Youth were given opportunities to decide on the topic, the content, and length of the movie and their role in the video production process. Video production was also a means for youth to tell stories about science and reconfigure their relationship with science and "to communicate on their own terms" (p. 670) their understandings of science. In turn, the mixing of text, images, and footage of special events, interviews with scientists, the public, or clips from the web led to multimodal meaning making in and with science. Building on our own work (Gonsalves, Rahm, & Carvalho, 2013), the video production process was essentially a tool to deconstruct science and engage youth in a dialogue about what counts as science and who gets to engage with it.

The nine participants were first-year high school students who volunteered to participate (7 girls; 2 boys; 13–14 years of age). They all came from the enriched science class. In practice, this meant that they were the kinds of students neither failing nor skipping school (interview of their teacher). Seven of the nine students were born outside of Canada (countries of origin: Tunisia, Morocco, Burkina Faso, Lebanon, Ireland, and Mexico). They worked on the documentary every week during one lunch period in their science classroom (60 minutes). Table 6.1 offers a summary of the video productions and the resources the club made available to youth or that were created by youth over time. It makes evident the complex cross-settings approach inherent in the design of the club.

In this chapter, I focus on the collaborative video production entitled *Impact of Invasive Species on Endangered Species in Urban Space*. Table 6.2 offers a brief summary of the activities the youth engaged in over the 12 weeks leading to that collaborative video production.

I analyzed the field notes and the video data of the 12 club meetings (10 hours of video footage total). I then pursued a content analysis, exploring the kinds of learning opportunities that emerged over time through the project.

Table 6.1 Summary of video productions, activities, and resources over time.

Time	Fall 2012	Winter 2013	Summer 2013	Spring 2014
Video Productions	1. Theory of the Big Bang 2. Biodegradable Plastic 3. Lynks & Sharks	Impact of Invasive Species on Endangered Species in Urban Space		
Resources Organized by Research Team	Visit by Scientist (Astronomy) Field trip in Community in Search of Science	Visit of Scientist (Environmental Sciences) Field trip to Green Open Space by River (adjacent community) Art Project	2 Week Internship Botanical Garden	Exhibit in Library: Art Work, Screening of Video Productions, Photography Exhibit
Youth Created Resources	Photographs of Field Trips	Progress Reports Calendar of Tasks Photographs of Field Trips		

Table 6.2 Summary of activities tied to the video production over time.

Time	Activities
Week 1	Brainstorm: What does biodiversity mean
Week 2	Field trip 1: Botanical Garden, Center of Biodiversity, & Insectarium
Week 3	Discuss field trip; develop topic of group video production
Week 4	Organization of production, subdivision into teams/themes
Week 5	Research on science content; calendar of tasks; storyboard
Week 6	Filming for documentary: Visit by scientist
Week 7	Video editing 1
Week 8	Video editing 2: Peer critique of video clip produced so far
Week 9	Video editing 3: Putting the three video clips of the teams together into one video, work on timing, fluidity, cut out long sections, add music, work on sound and transitions; watch & critique to improve
Week 10	Field trip 2: Visit a green space close to a river, close to school
Week 11	Video editing 4: Last day of editing work; final touches on video
Week 12	Video editing & sharing 5: Finalize documentary; watch it together

THE SPACE-TIME ORGANIZATION OF LEARNING AND IDENTITY IN/THROUGH A JOINT VIDEO PRODUCTION PROJECT IN SCIENCE

From Biodiversity to Invasive/Endangered Species in Urban Space

We started with a brainstorm session on biodiversity. All youth were asked to write something on the blackboard. It led to a list of terms such as ecosystems, diversity, biology, disappearance, nature, animals, plants, insects, microbes, and illnesses. The instructors then guided further brainstorming through a presentation and specific questions, exchanging with youth about biodiversity in terms of what it does for us humans and how it is endangered, and if so, what its consequences might be.

The following week was spent on developing skills in photography of objects of nature in anticipation of the upcoming field trip to the Botanical Garden, the Center of Biodiversity, and the Insectarium. Once on the field trip, youth were given digital cameras and asked to take pictures of things they saw that represented biodiversity for them. For instance, Ines noticed a banana tree with a bunch of small bananas, a discovery she shared immediately with Yousra and Sasha. Yet she was puzzled by their size. Instructor 1 then explained that there are many different kinds of banana trees, producing different kinds of bananas, including those we regularly eat. The banana tree was next to a carambola tree that Yousra also recognized. It prompted her to share her passion for that fruit to which Sasha replied by sharing her passion for its juice, noting, "that's even better!" It attests to the emotionally charged embodied learning youth could engage in as they observed and interacted with the natural environment in the greenhouses.

Youth also had many opportunities to make connections between their everyday practices and the various exhibits and objects they encountered. Their meaning making was spread out spatially across practices. Once in the Center of Biodiversity and asked by their guide of the center for a definition of biodiversity, Maarika responded quickly, to the surprise of the guide, "it's the diversity of life." She could reinvest what they had discussed previously in the club. The following week, in teams of two, youth prepared a PowerPoint presentation with six pictures from the field trip that best represented biodiversity for them. They struggled with speaking in front of their peers, which in the end was good preparation for the kind of talk they needed to engage in either with scientists or for documenting science in the video production.

The objective of the third week was to settle on a topic for the video. The discussions about biodiversity set the background for the video to be developed. Yet the two instructors challenged youth to be more specific and think of a storyline for their documentary that would end with a message for community action. Marc, one of the youth, wanted to focus on the ecosystem because "everything is related to the ecosystem somehow." That was seen as too vast a topic and youth were encouraged to be more specific. Mellan suggested "pandas or bamboo forests." The second instructor encouraged youth to choose a theme more closely linked to their lives and communities. That prompted Yousra to suggest "urban ecosystem, invasive species, endangered species," which prompted Lydie to add "illnesses" and Marc "humans." Yousra approved it by noting "that's it, we keep the best for the end!" It made the group settle on "the urban ecosystem." Instructor 1 then encouraged youth to "narrow down and define the theme a little bit more." Yousra offered a narrative . . . : "In the urban ecosystem, there are invasive species like the pigeon, the squirrels, the cat, and well." It led to a group discussion tied to the video on biodegradable plastic that they had created in the fall. Then instructor 2 reoriented the discussion by noting, "We could start with the impact of invasive species in the urban context," which was received positively by the group.

The next challenge was to find ways to address the topic and develop a storyline. Instructor 1 explained that there are really two dimensions to the work of scientists; on the one hand they study the theme, but then they also examine the problems associated with the theme and how the problems might be resolved through concrete actions. Instructor 2 reframed the task by noting, "First we have to further refine our theme but then also identify an action that it leads to or calls for." Later that day, the instructors introduced another activity in hopes to further scaffold the development of the storyline of the movie. They proposed a link between biodiversity and apples they were eating. Instructor 1 asked youth about a link they can make between the apple and biodiversity. Yousra suggested, "The worms inside!" Instructor 1 responded, "But most apples do not have any worms inside, do you know why?" Alice responded, "because of pesticide." She explained how pesticide is used to ensure the growth of the trees and apples that can be sold. Instructor 1 asked, "Do you think the pesticide affects only the apples?" Marc offered, "It affects the air as well," while his peers were silent. It led instructor 1 to go into more details about the connection between pesticides and the overall ecosystem and the ways pesticide affects the apples but also insects, animals, plants, and other things that are alive. Marc suggested to simply no longer use pesticide and to buy biological products. Mellan, added, "Or just plant apple trees inside!" The two instructors were somewhat surprised by that response and instructor 1 reminded the youth that greenhouses actually consume a lot of energy and that insects still enter anyway. Instructor 1 expanded upon Marc's suggestion about buying biological apples and explained that biological farms also have insects, which are then controlled through the introduction of certain kinds of birds, such as the bluebird who nourishes itself on insects to balance the two species.

The tracing of the first four weeks of video production makes evident the kinds of learning opportunities and identity work the activities supported over time. That process was further mediated through social interactions and tools that were sought out, designed, or brought in. I now examine more closely some of the identity work the video production process supported.

Development of Creator Identities Through Remix and Creative Productions

> I enjoyed the collaborative project. That way, everybody had some ideas and we could all talk to each other and discuss things. Everybody had a task and had a responsibility in it. It was like OUR documentary. That's why I enjoyed it.
>
> (Yousra, Interview 2013)

The video production called for a team effort, while it also offered much space for youth voice, a balance that was not always trivial. First, the overall theme of the video was divided into four manageable subthemes in week four, with Marc and Mellan assuming "the different urban ecosystems," Lydie and Sasha "endangered species," Yousra and Ines "the relationship between the urban ecosystem and endangered species," and Alice, Laura, and Maarika "invasive species in Montreal." A calendar was also developed and then taken over by youth, indicating each week the specific tasks and objectives were to be met by the end of the session. It led youth to realize that they had little time left for the actual editing of the movie, putting much pressure on them to work more efficiently. Each subgroup then worked on a storyboard for their part of the production and in light of the scientific research they had pursued previously. We arrived with the "official" storyboard sheets with aligned squares for a description of the chosen pictures or video clips and spaces below for the addition of the chosen sound, special effects, timing, etc. However, it became clear rather quickly that the planning of the video had to be pursued within less rigid structures, and teams of two were given large poster boards that could then be moved and reorganized as needed once the sequence of the documentary became clearer.

The following week (week six), different youth representing the subthemes of the planned documentary—introduction, urban ecosystem, endangered species, and invasive species and conclusion: solutions to problems—had prepared questions for the visiting scientist. They filmed those exchanges and used either the content in their video production or parts of the actual video footage. The following week, youth started editing. At the end of the session, each clip was viewed and feedback offered. Instructor 1 opened up the floor for questions. Kathia felt that some information needed to be added, such as a note to the overall theme prior to the interview, for instance. Maarika called out, "We call that an introduction!" Alice then challenged the team on the duration of the interview noting "Fifty-four seconds of the interview is a bit long." It led to a discussion about documentaries and how some of them do contain long segments of interviews but not others. Feedback by the whole group was sought on three occasions as the editing progressed. It led to heated debates at times, as the following exchange suggests with Yousra noting "We don't write enough, we rely too much on the interview" and Marc suggesting, "The images go by way too quickly, I did not have time to read the text" and Alice, "I find it a bit long, maybe too many clips from the interview, not enough things to read!" While not trivial, the mixing and creative blending of different cultural artifacts is illustrative of the kind of multimodality video editing called for and made possible. Meanings and selves were articulated and stretched across the multiple modes such as music, sound, image, gesture, space, moving images, and a blend of those modes and synchronization of the visual with the audio over time.

The story was eventually told through the actual movement and pace of the film, a rather challenging task. For instance, Lydie was focused on the content they were working on, "We are doing invasive species, right?", whereas Yousra focused on the manner they would transmit the content to the viewers, "We will write a short summary." The latter was challenged by Lydie "Again?" who then counted the summaries they had so far, arriving at six. Ines was wondering about the length of the clip so far and the challenge a lengthy text posed for the actual duration of the clip, dimensions they struggled with at the same time. The exchanges centered on what to include but also the timing and duration of the scrolling texts. It was quite challenging for youth to work effectively as teams while also developing their technology expertise in video editing. It led to the development of an identity as creative producers with technology. As Alice noted at the end of the school year, "I am now better with computers, I used to be such a dork" or Marc, who explained how the project made him "communicate (talk) more and better with others."

The final cut begins with a clip that was shot by the media team at the school. In that part, Yousra and Marina briefly described what the club was about and why they participated. The girls positioned themselves as liking science and as enjoying the forms of engagement with science the club supported. Yousra noted, "We like it," and Marina added, "We like science," which led Yousra to restate their interest in other ways, "We like to know more about science," which made them laugh but insist, "It's not a joke!"

As shown in Figure 6.1, the final cut implied the remixing of images and clips taken from the web and their field trips, interwoven with rolling scripts and summaries of science content tied to the theme, next to short segments of the interview with the scientist, carefully selected to make for an engaging storyline.

A SPACE-TIME ANALYSIS OF LEARNER-ENVIRONMENT RELATIONSHIPS

As summarized in Tables 6.1 and 6.2, engagement in the club activities also implied spatial travel to science venues beyond the immediate classroom. The design of the club was grounded in a cross-setting approach to learning and identity development. The Botanical Garden, the Center of Biodiversity,

School A: *The Impact of Invasive Species on Threatened Species in Urban Space*
Excerpt below 2:15min (total length 7:25min); all along same electronic music typical of science documentaries; text (at times multiple ones following one another) & scientist offer explanations and science content.

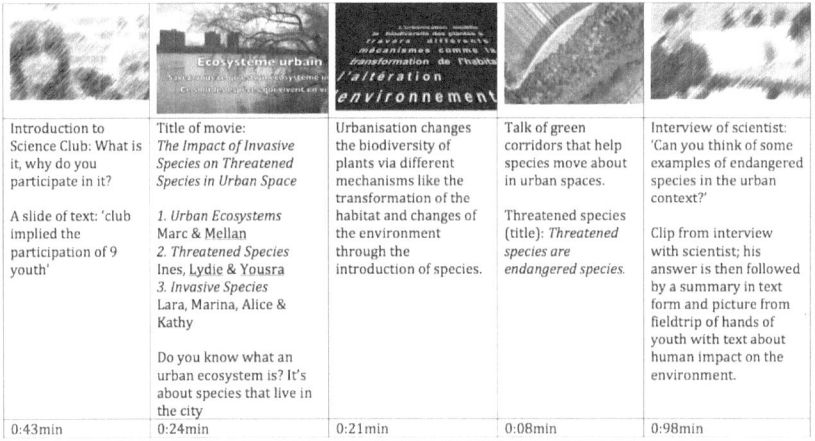

Introduction to Science Club: What is it, why do you participate in it? A slide of text: 'club implied the participation of 9 youth'	Title of movie: *The Impact of Invasive Species on Threatened Species in Urban Space* *1. Urban Ecosystems* Marc & Mellan *2. Threatened Species* Ines, Lydie & Yousra *3. Invasive Species* Lara, Marina, Alice & Kathy Do you know what an urban ecosystem is? It's about species that live in the city	Urbanisation changes the biodiversity of plants via different mechanisms like the transformation of the habitat and changes of the environment through the introduction of species.	Talk of green corridors that help species move about in urban spaces. Threatened species (title): *Threatened species are endangered species.*	Interview of scientist: 'Can you think of some examples of endangered species in the urban context?' Clip from interview with scientist; his answer is then followed by a summary in text form and picture from fieldtrip of hands of youth with text about human impact on the environment.
0:43min	0:24min	0:21min	0:08min	0:98min

Figure 6.1   Content analysis of a short segment of the final documentary.

the Insectarium, and an open space area in an adjacent community to the school were the kinds of venues solicited to ground students' understanding of science in authentic settings. Youth were given opportunities to observe, interact, manipulate, touch, and smell objects of science in place. Such forms of engagement were supportive of the development of emotionally charged ways of knowing and being in science and in learning environments new to youth. Take, for instance, our field trip to an open space in an adjacent community and close to a river in week ten. That field trip served in part to reengage the youth with their video production. Yet most importantly, it led to embodied understandings of issues tied to biodiversity in the city that were emotionally charged and grounded. For instance, instructor 1 made Alice think of her video documentary on pollution in light of the waste apparent in the park, asking Alice whether she thought it had an impact on the animals in the park. Alice shared her observation of a bird nest that was made with dog muck. Kathia added, "The bird did an excellent job integrating wastes into the nest, there was even a string," a piece of rope from boats that the birds possibly found in the river. Kathia argued that the string attached the nest to the tree "with a sort of a knot" to which Marc added, "Very smart."

As shown, instructor 1 guided youths' observations and helped them make connections across space and time by suggesting relations between what they were "seeing" and what they had researched previously in the context of a video on biodegradable plastic and waste in the fall. Kathia's response is also an excellent example of *pereživanie* in that her comment makes evident not simply a meaningful observation and cross-setting connection of science content in light of her previous and current video production but also its emotionally charged nature. She was impressed by the skill of the bird in building a nest by combining waste materials with natural waste. It resulted in a functional nest, something Marc also admired. Kathia also took a picture of the nest to share with her peers later on. That picture was then reused in their arts project as well. That field trip led to embodied and emotionally significant learning according to instructor 1:

> It was pretty effective, because we were talking about pollution in urban areas and were able to see a bunch of things first hand, and were able to see bird nests with garbage incorporated in them, and I think that had a much bigger effect on the kids than talking about it.

The weaving together of work in the club with nature helped youth "see" nature in the urban context in new ways and appreciate what a complex system it implies. Another example of the manner thought and emotions were interconnected and constitutive of learning and becoming within the club is evident from our visit to the Botanical Garden, as a comment by Yousra makes evident:

> My mother encouraged me to go on the fieldtrip and see with my own eyes instead of just talking about it in class . . . the Botanical Garden, it was amazing, I felt like being in a real forest, yet I was still in the city!

The field trips and instructors guided youths' gaze at their environment, putting to work the science knowledge they had constructed together previously through the video production. Maybe not surprisingly, the club made science available to youth in a different way than their school, as Alice suggested, "In science class we always do worksheets, each time on something different, in the club we did projects where many activities and topics were related."

Engagement with science was also embedded in productive and emotionally safe and supportive relationships among youth and the club instructors and tangled up in rich relationships with a multitude of learning environments. As one youth noted,

> I really like science since I am a small child, but what I liked the most in the club is doing science and all the activities without constraints, without constantly being looked at by others, being noticed by others, there was more freedom to just do science.

According to instructor 1, the club also helped youth "to think about bigger subjects, and just taking the time to think about different issues, be more open-minded about stuff and critical, ask questions." It was a safe space to explore science in ways meaningful to youth and engage with content at a deeper level given the spatial arrangement of learning that was stretched across science venues and over time. The club was youth centered, supportive of youth voice and agency, and, as such, charged with a climate grounded in positive emotions.

THE VALUE OF A SPACE-TIME LENS FOR UNDERSTANDING AND DESIGNING LEARNING OPPORTUNITIES THAT MATTER

By tracing the video production project across space and over time in this chapter, I show how opportunities for learning and identity work were jointly organized by the participants and instructors and emerged from ongoing dialogue that centered on weaving together different ways of knowing and being in science. That weaving together also implied spatial travel in that youth moved back in time and made connections across space, linking prior understandings of science gathered in school or at home with those being developed. Spatially, youths' learning and becoming was distributed among informal science venues that were sought out on purpose to contextualize in different ways the science topics that we explored together. That kind of spatial travel helped youth contextualize abstract scientific knowledge they had gathered through their research in ways meaningful to them. It helped them look at their urban environment in new ways as the noticing of the bird nest by youth suggests. It made them develop an understanding of their urban ecosystem by engaging with it physically, mentally, and affectively. The study confirms that learning and identity need to be understood holistically and locally, in light of the learning environments in which they take shape, yet also globally and as tied to and made through youths' mobility among learning environments and cross-setting meaning making.

The club also fostered the social and emotional capacities of its participants, given its cross-setting design and grounding in supportive relationships between the youth and their instructors as well as places of learning next to the tools that were made available or created by youth. Those relationships then sustained the learning and becoming of youth as creative agents of their ways of knowing, being, and video productions. It made possible the kind of holistic learning Vygotsky talked about, offering insights into the kinds of caring actions, communication, and dialogue that are at the heart of a space that "redresses the bifurcation of cognition and emotion in education" (p. 222). In addition, photography and video were tools supportive of a move toward participatory research, giving voice to youth and positioning them as key contributors to the ongoing research project. These tools helped youth represent their understandings but also led youth to more deeply engage with biodiversity and its many dimensions. Youth created tools such as their adaptation of the storyboard practice or identified ways of dividing labor during the video production that were also constitutive of a practice grounded in youth voice and driven by holistic learning and empowering ways of becoming in science. Hence youths' learning and authoring of selves were tied up in complex ways with the kinds of resources and tools youth had at their disposition, sought out, and appropriated under guidance by the instructors. That way, the club also made accessible to youth "identities as doers and learners of those practices" and of and in science (Nasir, 2012, p. 33). It led to the emergence of a practice grounded in creative acts that "promoted new connections among learning opportunities inside and outside of school" (Barron, Gomez, Pinkard, & Martin, 2014, p. 42) and among cognition and emotion in ways captured by Vygotsky's notion of *pereživanie*.

After-school and community science programs are a means to make quality science accessible to youth in ways complementary to schooling. They can be particularly important for youth from underserved communities who lack access to quality science elsewhere. Yet to offer rich science learning opportunities in after school is "a tall order that will require positive changes among staff, sites, and at the policy and funding levels" (House, Llorente, Gorges, Lundh, & Mata, 2015, p. 38). As is, few after-school science programs are project driven or engage youth in inquiry and/or deep science in ways described here, over extensive periods of time. In this chapter, I make the case for a cross-setting lens and complex space-time driven approach to forms of engagement with and identity work in science, at the heart of a science practice driven by youths' interests and emergent from rich relationships with others and place. While challenging to implement, we see the kind of creative partnerships at the heart of the Fifth Dimension Model and entailing affinities among universities, schools, and the community, as one way to move toward that ideal—which is essentially about complex intersections of formal and informal science practices.

ACKNOWLEDGMENTS

I thank the animators, students, and research assistants for their contributions to the club (Émilie Boulanger, Issac Hebert, Gwénaëlle Journet, and Audrey Lachaîne), a team project that led me to write this chapter in the plural voice. I also want to thank all the youth, their parents, community organizations, and the school. The study was supported financially by Fonds de recherche, Société et culture, du Québec (FQRSC).

REFERENCES

Barron, B., Gomez, K., Pinkard, N., & Martin, C. K. (2014). *The digital youth network. Cultivating digital media citizenship in urban communities.* Cambridge, MA: MIT Press.

Cole, M. (and the Distributed Literacy Consortium) (2006). *The fifth dimension: An afterschool program built on diversity.* New York, NY: Russell Sage Foundation.

Furman, M., & Calabrese Barton, A. (2006). Capturing urban student voices in the creation of a science mini-documentary. *Journal of Research in Science Teaching, 43,* 667–694.

Gonsalves, A., Rahm, J., & Carvalho, A. (2013). "We could think of things that could be science": Girls' refiguring of science and self in an out-of-school-time club. *Journal of Research in Science Teaching, 50,* 1068–1097.

House, A., Llorente, C., Gorges, T., Lundh, P., & Mata, W. (2015). Science in California's public afterschool program: Exploring offerings and opportunities. *The Journal of Expanded Learning Opportunities, 1,* 31–39.

Nasir, N. (2012). *Racialized identities: Race and achievement for African-American youth.* Stanford, CA: Stanford University Press.

Vygotsky, L. S. (1987). *The collected works of L. S. Vygotsky Volume 1: Problems of general psychology.* New York, NY: Plenum.

Vygotsky, L. S. (2004). Imagination and creativity in childhood. *Journal of Russian and East European Psychology, 42*(1), 7–97.

7　When Science Is Someone Else's World

Emily Dawson

> It is quite confusing I think, I don't know, I like the idea of liking the [science] things, but actually to do them is something else. So I like the sense of going to the zoo, purely just to see the animals, but I wouldn't. I don't like touching them. I don't like being in that kind of environment. So I think it's all about fascination. So you said I like how things are built and how things are, so seeing an animal in a cage, it gets my brain thinking, oh the journey that the animal had, like to get a lion from the jungle, to get it to the zoo, the cage, and see how the lion actually adapts, 'cos it's a different scenario for the lion. So things like that are interesting for me, but the thought of touching animals, it's not my cup of tea, in terms of the museums and stuff like that, I, it's not my cup of tea at all (laughs).
>
> Fatima, Somali Group

How does it feel when informal science education (ISE) isn't your "cup of tea?" I first met Fatima in February 2010 while exploring social exclusion from public science and ISE. There are two key things you should know about Fatima for this chapter; first, she loved particular aspects of science and, second, despite her support for the study, Fatima was an unwavering critic of ISE.

As a former ISE practitioner, I knew it was as easy as looking across a crowded gallery to know ISE was not very inclusive. Surprisingly, at that time, little research was available about how exclusion from public science worked, let alone how ISE could become more inclusive. My colleagues and I found that job titles like "community officer" and "diversity manager" were sometimes used by our institutions to partition equity issues off from day-to-day work, often against the best intentions of those involved. As Ahmed (2012) notes in her study of diversity workers in higher education, institutional practices can limit such roles; creating inclusive job titles can give the appearance of inclusion, while making little structural change. Faced with these mounting frustrations, I set out to explore ISE from a different perspective, that of a nonvisitor.

In this chapter, I draw on data from what became an ethnographic study carried out with four grassroots community groups in London to explore how they engaged (or not) with science in their lives. Over two years I worked with a Sierra Leonean group, a Somali group, a Latin American group, and an Asian group, attending group meetings (dances, festivals, celebrations, picnics, and ISE visits) where I carried out interviews, focus groups, and observations. The groups were approached on the basis of data that showed ISE participation in the UK was marked by race/ethnicity and class (Dawson, 2014a). All participants therefore came from low-income, minority ethnic backgrounds, but a range of different ages.

I begin this chapter by outlining why it is important to consider public science through the lens of equity and the theoretical tools I use to do so. I then briefly describe attitudes and experiences of science across all four groups, followed by a more detailed account from one person, Fatima. In looking at Fatima's stories and experiences, I hope to illustrate that being interested in science does necessarily pave the way to participation in public science activities such as ISE. I argue that exclusion from public science is not a question of rebranding and changing perceptions, but instead goes to the core of how ISE is understood and practiced.

WHY DOES INCLUSION IN SCIENCE MATTER?

Where, how, with whom, how much, and why we engage with science (or not) matters. I frame group and individual experiences of science, ISE, and ISL (informal science learning) against the social reproduction of disadvantage because it is against that backdrop that questions of inclusion and exclusion are important. If science were irrelevant, it would not matter who spent their time among its institutional norms and texts, absorbing the language, shaping the appropriate ways of being, or imagining themselves in future science stories. Science is a prized resource in our societies. It is therefore important to map where people encounter science in their lives and what happens when they do.

Public science takes many forms, from the overtly political to activities designed purely for fun. In this chapter, I focus on science in general as well as science learning in designed, institutional spaces (ISE)—museums, science centers, zoos, or science festivals and more—and the harder to map informal science learning (ISL) that happens every day. The ever-growing field of ISE institutions is potentially a useful space for people to engage with science, to imagine themselves within the world of science, and to try out being science insiders. Alongside these institutionally structured practices are millions of more nebulous science encounters in the wild (ISL), including reading science stories in newspapers, following science stars such as Neil deGrasse Tyson on twitter, watching science on television (from *The Big Bang Theory* sitcom to the Planet Earth documentaries), or chatting about science among friends and families. I focus on both ISE and ISL here because they infiltrate people's lives in different ways.

Does this wide and varied field of public science practice create multiple, equitable pathways around, through, or into science for everyone? Unfortunately not. For example, ISE practices appear to be exclusive, marked by social structures such as ethnicity, class, gender, and other social positions (Dawson, 2014b). This means that exclusion from ISE is hierarchical (because it reflects patterns of social disadvantage) and intersectional (because those disadvantages overlap). But perhaps we should not be surprised that public science activities are exclusive, because school science is also patterned by privilege. Research in science education has shown that some people get turned off of science at school, whereas others are supported to pursue science studies and careers (Brown et al., 2015). What then might it look like if this vast array of ISE, ISL, and formal science learning opportunities were not for you? How might it feel to be excluded?

THEORETICAL BACKGROUND

To understand how a sense of being excluded from science might develop and be reproduced, I draw on concepts from research on social reproduction and exclusion. In doing so, I stray into the dubious territory of describing people as excluded, what Becker (1963) called "the act of labeling, as carried out by moral entrepreneurs" (p. 179). It is important, therefore, to note that research is as guilty of reifying social divisions as any other practice. I explore how people and power come together in potentially damaging practices, with a view to describing and changing that system.

I use Bourdieu and Passeron's (1990) work on symbolic violence to think about the relationships between a specific field (public science, ISE, ISL), the forms of capital valued by that field, and the disposition toward the field, or habitus, of those involved. Symbolic violence is the misrecognition of power and agency, such that the disenfranchised—the working class—make a virtue of necessity by interpreting inaccessible opportunities as choices not to participate. In other words, symbolic violence is present when exclusion from a given field of practice or set of institutions feels like something so anticipated by your ways of thinking that you might never expect to be included, that your exclusion feels natural, and, sometimes, desirable.

Imagine for a second the unthought assumptions that guide your day-to-day life. I, for example, automatically walk into female, not male public toilets, I sit upstairs but not at the back of London buses, and I avoid unlit parks at night. All these embodied practices emerge at the junction of who I am and how I understand my places and roles in the society I live in. From this perspective, not using ISE or disliking science could become an embodied disposition, a way of being, developed across groups whose experiences are similar, habitus.

Institutions are renowned mechanisms of social reproduction. As such, we need to pay attention to questions of belonging, who feels welcome and unwelcome in science, ISE, and ISL, what Ahmed (2012) describes as "how some more than others will be at home in institutions that assume certain bodies as their norm" (p. 3). Thus, in this chapter, I use the device of insider/outsider as a way to think through what might be involved in participants' experiences of science and how they position themselves in relation to science, ISE, and ISL as a result.

I locate insider/outsider positions within participant's identity practices to frame identity practices as fluid, reimagined, or reinforced in specific contexts and rooted in relationships with others, although enacted by individuals. In contrast to the notion of collective dispositions, or habitus (Bourdieu, 1998), the notion of identity in practice helps me to think through the differences between people, as well as where they may be similar. Unfolding identity practices at an individual level, therefore, means looking at how ways of being, learning, and becoming are traced through with historic, social, and political features but remain open to change and agency. In this crucial sense Holland, Skinner, Lachiotte, and Cain (2001) leave agency foregrounded in their understanding of people's actions in ways that Bourdieu's work is less attuned to. Whereas it is vital to unfold how power and exclusion operated in science and ISE, so too must room be left for people to genuinely reject participation in ISE, even if *at the same time* the conditions are such that they would be excluded anyway.

SCIENCE AS INACCESSIBLE AND UNAPPEALING

On Being Disposed Against Science

"Science . . . it's a subject very far from my reality, from what I do," stated Alejandro from the Latin American group. Like other participants, science was something Alejandro felt he had no control over, no stake in, and could not imagine a scenario where he might be more involved in science, whether politically, culturally, socially, or educationally framed. Across the four community groups, participants described an overwhelming disassociation from science, at school, ISL and ISE settings, jobs, or any other aspects of their lives where they thought they might encounter science.

This sense of alienation, of being outside or tangential to science and public science was particularly acute when it came to ISE. With few exceptions, science museums and centers were unfamiliar to participants and those visited as part of the study highlighted how exclusion was embedded in ISE practices such that the visits confirmed their preexisting views of ISE as problematic and exclusive (Dawson, 2014b). Where participants did have experiences of ISE to reflect upon, such spaces were described as whitewashed and Eurocentric, expensive, irrelevant to their lives and communities, and, as a result, worth avoiding.

Participants saw ISE institutions as unwelcoming, hostile places, where they did not belong, drawing on their perceptions of institutional whiteness and their sense of being outside ISE and outside science. Being outside exists only in relation to the possibility that someone is inside. We should recognize therefore, as Ahmed (2012) reminds us, that the problem of exclusion is not that of *perceptions* of institutional whiteness but of institutional whiteness itself—ISE practitioners and users in the UK are drawn overwhelmingly from the white ethnic majority (Dawson, 2014a).

In contrast to ISE, participants in all groups remarked on ISL encounters, particularly watching science on television (from detective shows to comedies). Such programs were however, not framed by participants as supportive of an orientation toward science or of being science insiders; for all that, they were perceived as more accessible. Watched for entertainment value, science on television featured an all-star cast of people who were "not like us." As Kirin from the Asian group put it when talking about the television series *CSI* (Crime Scene Investigation), "we're very interested but you know, we can't push ourselves forward." For her, her friends, and participants from other groups, science on television was represented by people who were special, impossibly clever, but not like them, and they did things they could not do, echoing the idea found repeatedly in science education that science was difficult and the reserve of the "genius" few. In the same breath, therefore, as participants named a series of famous white male science presenters, including Sir David Attenborough, Steve Irwin, and Sir Patrick Moore, they highlighted the social distance between themselves and their perception of who was involved with science.

Across all four groups, science was perceived as a difficult and unpleasant subject to study, of little relevance, little interest, and little use to participants or their communities. As Maria, a mother of four from the Latin American group explained, "The way science is presented at school is very boring and uninspiring." Concerns about employability and income influenced many participants views on whether pursuing an interest in science was worthwhile. Formal science education appeared especially irrelevant, because pursuing a scientific job was seen as impossibly hard, backed up by stories of friends and family who had tried and failed to work in the sciences. In each community group, participants (with three exceptions, one of whom was Fatima) talked about how they had stopped studying science as soon as they could, with some specifically noting school as the key factor that had put them off science.

The collective sense of disassociation from science and ISE across the four groups is striking in terms of habitus and symbolic violence (Bourdieu & Passeron, 1990). Science was a difficult subject, off-putting at school, and of little value for work, whereas ISE appeared invisible, pointless, and exclusive or, in the case of ISL and television, entertaining but not something they could identify with. Science was understood to be marked by race/ethnicity, class, gender, and, in some cases, age, in ways that did not welcome participants.

As a result, participants steered clear of public science activities, withdrawing from a system they interpreted as disadvantageous and arranged against their interests. They were disposed *against* science and public science. As Bourdieu and Passeron (1990) argue, the most effective form of domination is that which "comes from exclusion, which perhaps has the most symbolic force when it assumes the guise of self-exclusion" (pp. 41–42). Participants saw themselves as science outsiders and behaved accordingly. For participants, science was historically, socially, and culturally constructed as a world for people who were, as Mirza from the Asian group concluded, "not like us." In this sense their involvement was framed as hard to imagine, unwanted, unthinkable, and unlikely.

Fatima: When an Interest in Science is Not Enough

When I met Fatima she was in her mid-20s and had been involved with the Somali community group for several years. She became a key participant from that group; supporting the research as a gatekeeper and general explainer who unpicked the nuances I missed, translating (literally and conceptually) between other members of the community group and me. Exploring Fatima's experiences and attitudes is interesting because among the 60 people who participated in this study, she and two others were the only ones who expressed personally liking science and had tried to study it further, albeit without success. I discuss Fatima's stories, reflections, and experiences here to show how some of the themes briefly sketched earlier appeared in the context of someone's life.

Fatima had grown up in the UK, going through the British school system and to a post-16 college, although not to university. She lived at home with her extended family of siblings, their spouses, their children, and her mother. Fatima described herself as the "odd one out of the family, I'm a weirdo" because she was interested in science and preferred staying in reading books to going out but agreed that "none of us really like museums." In talking passionately about science one day she said, "I've got a fascination with biology, how the body functions and how each part of the body has a function." It turned out that the kinds of books Fatima read were also unusual from her perspective compared with what her friends and family enjoyed reading; Fatima read science books—specifically books on engineering and biology. It seems safe to say that Fatima really liked science.

Not only was Fatima into science and reading books about it, she pursued her interests in science through other forms of ISL, seeking out ways to develop herself in relation to science purposefully through specific practices. For instance, she talked about going online to research her scientific interests and being known among family and friends as good at finding useful scientific information when it was needed. Unlike other participants, she chose to watch television programs with a lot of science content. For instance, in the following extract, she describes a nature documentary she had enjoyed:

> I'm fascinated just to look at an animal and then to see how the animal came about, 'cos like, I was watching a documentary the other day and like, um, small animals like birds when they were tiny, they were showing how they develop, and how in a couple of months they get bigger and bigger and you have the big pigeon that you have.

Fatima's presentation of herself and the views held of her by others that she echoed were at least partially built around her orientation toward science, her skill with scientific information, and her seemingly well-known like of science. In other words, science featured in her "practiced identity" (Holland et al., 2001, p. 271).

Being disposed toward science was, for Fatima, not as straightforward as a habitus that endowed her with a "feel for the game" of science and public science (Bourdieu, 1998, p. 25). It was with some discomfort, some residual sense of outsider status that Fatima positioned herself as different to her family and friends through her unusual or "weird" interests in science and her choice of ISL activities, as though she was misaligned with the collective habitus, the collective disposition against science within her community. In this sense, Fatima saw science, scientists, and those with science interests like herself as unlike other people, echoing the statements of other participants and other studies about scientists as "geniuses" (Lemke, 1990).

In conversation with her friend Idyl, another Somali participant in her mid-twenties, Fatima described scientists as different, agreeing that they were not "normal" people like her friends. In these conversations, scientists appeared compelled to further science, no matter what the social or ethical costs, morally dubious and alarmingly clever, outside the social norms and behaviors she expected. Fatima struggled therefore to negotiate her disposition toward science, or habitus, using ways of talking, behaving, and other identity practices to bridge perceived social distances and deviant behaviors between herself, her community, science, and scientists. Thus in how she presented stories of herself in relation to science writ large, Fatima worked to balance a self that was both science insider and outsider, both "weird" and "normal."

On "Hating" ISE

In one of our first meetings, Fatima bluntly told me, "I hate museums." In a later interview, she continued, "I'm very upset with the museums, so I'm not going . . . I just did it because I had to do it at school, but now it's not part of my social outlook, why do something you don't, it's not part of you." Compared with her presentation of self in relation to science in general terms, Fatima saw herself as definitively outside ISE but also, crucially, that

ISE was outside and irrelevant to her life and her friends, as she said, "not part of you." Fatima's experiences of and attitudes toward ISE were in line with those of the other participants. That is to say, with a collective habitus or disposition that oriented them away from ISE as by and large unheard of, unusual, unhelpful at best, and damaging at worst.

Unlike most of the other people involved in the project, Fatima was able to draw on her previous experiences of ISE at length, because she had visited several museums and similar institutions, including the Natural History Museum, the British Museum, the Science Museum, London Zoo, and Vauxhall City Farm. But these visits did not mean Fatima liked ISE. On the contrary, she told me she thought ISE outreach practices failed to meet the needs of those they should (all of the public) and were simply not up to standard. She said she had never seen an advert for an ISE institution in her neighborhood, nor leaflets, signs, or information in community newspapers, websites, or on radio stations and that she felt her community had been left to one side as a result.

Fatima's views of ISE institutions had been influenced by her experiences while at school. In fact, all but one of Fatima's ISE experiences (Vauxhall City Farm) had been via a school trip. Fatima described school ISE visits in negative terms as "a sort of detention" and "punishment." These experiences were strongly framed by what Holland et al. (2001) call a "figured world" within and against which people's identities develop, in this case, the figured world of compulsory schooling.

> It wasn't enjoyable, any museum that you went to as a kid, it's not really what you would say is the best time that you had in school, like it was like a punishment, they would say that you're going on a trip, and then you turn up at the Natural History Museum or the Science Museum or something like that, and it's not really a trip that a child imagines, you know, the night before, packing it's pack lunch, you don't really imagine that you'll be in a tour with a tour guide that doesn't really care because it's been doing it all day and you're the last group, and it's like whistling through the whole museum, so, in terms of that, no I wouldn't, I wouldn't imagine putting myself in that.

In the extract, Fatima contrasts childish delight at the idea of a day away from school with the disappointing realization that ISE mimicked the figured world of school, complete with teachers, bored guides, and rules to follow. ISE visits were motivated from Fatima's perspective by her teachers' seemingly incomprehensible love of ISE spaces, with largely undelivered potential for learning science and having fun. So much so, that once ISE visiting was no longer mandatory, Fatima had tried never to visit them again.

Fatima's underlying assumption was that participation in ISE or anything like it would be unusual for her, her friends, family, and broader community. As she put it, "I don't know anyone that's decided one day 'oh, let's go to the museum.'" Her experiences generated a story about ISE as poor-quality, off-putting, and irrelevant, a story reinforced and reproduced socially among her community and friends into a world where science and ISE was inaccessible, unpleasant, and removed from day-to-day life.

Despite her personal interest in science, therefore, visiting an ISE institution was not a choice Fatima expected to make, nor did she expect her friends or family to do so, as the following extract shows:

> FATIMA: If you don't know anything about the museum and it's not part of your social outlook then you don't know what's happening in the museum.
> EMILY: Yeah, and you'd never look it up?
> FATIMA: You'd never look it up, you wouldn't have no need to because it's not something you do.

This extract speaks to a deeply ingrained sense of not belonging in ISE, but it also echoes how Fatima made sense of her alienation from ISE; she was an ISE outsider, but ISE was in turn outside her life and her community.

Fatima's stories about science and ISE provide a useful account of the complexity of people's lives and a concrete context for exclusion from science, ISE, and ISL. Fatima shared the collective disposition against ISE with other participants, but unlike others, she was disposed toward science and certain ISL practices. She worked hard to find ways to understand what at times felt like contradictory dispositions. Holland et al. (2001) have suggested that people frequently face situations of contradiction, where one or more aspects of their positional identities conflicts with another aspect. Similarly, Roth (2008) has argued that engaging with science from a marginalized social position creates cross-cultural differences that require considerable negotiation and produce multiple, heterogeneous identities. As Fatima said in the quote that opened this chapter, "it is quite confusing" to both like science and dislike certain kinds of science engagement opportunities.

Fatima's descriptions of the irrelevance of ISE to her life and her confusing social distance from science can be interpreted as an articulation of her experiences of marginalization and as a way to resist such experiences. Through constructing positions from which to criticize ISE practices, Fatima was able to acknowledge the ways in which she was excluded from such practices, while simultaneously rejecting those practices on her own terms. Fatima did not like science because of her ISE experiences, but rather in spite of them.

Understanding how Fatima made sense of her views and experiences of science through contradiction, confusion, being a science insider, but an ISE outsider suggests that agency and identity work play a key role in negotiating between individuals, fields, and collective habitus. Disidentification with educational institutions can be considered a form of agency. Drawing on Holland et al. (2001), I suggest in addition to being structurally excluded, through their behaviors and speech, participants actively spurned science and public science. Thus whereas people's individual positions toward different aspects of public science varied—Fatima rejected ISE whereas other participants disidentified with public science activities and science altogether—they were not wholly passive in their exclusion.

In this sense, agentic rejection and structural exclusion go hand in hand to reproduce social disadvantage; with both sides of the coin in play, exclusion/rejection becomes a resilient system, hard to change, and rooted in symbolic violence (Bourdieu & Passeron, 1990). Symbolic violence then is "based on 'collective expectations' or socially inculcated beliefs" (Bourdieu, 1998, p. 103). It sneaks into well-meant intentions, in doing what you have to do, usually what you do and expect to do. As Fatima put it, "why do something you don't, it's not part of you."

CONCLUSIONS

This study taught me that exclusion and rejection are habits of mind, embodied practices, assumptions, and expectations that work together. Participants' disassociation from science in general and ISE in particular made their exclusion all the more resilient. That participants' expectations of being ISE outsiders were met in practice by those who visited museums and science centers as part of the study was even more appalling (Dawson, 2014b). Their exclusion was embedded in the ISE practices they encountered, written into exhibit texts, and mirrored in photographs. The ISEs participants visited were whitewashed, not only in terms of the other people there, but in content and representation; people who looked like them were either invisible or the subjects of science, stars of exhibits about evolution or disease, but rarely (if ever) the revered scientists themselves. The pernicious combination of people being disposed against science and ISE, rejecting a system that disadvantaged them and their structural, institutionalized exclusion created a world where participation in science activities was marked by privilege in ways that were durable.

The resilience of the assumption that science and ISE are for some but not for others is the hallmark of symbolic violence and the reproduction of disadvantage. If we are going to take seriously the challenge of making science inclusive, we have to disrupt these expectations and beliefs at their roots. Fatima's stories show that we must unsettle the idea that people do not participate in ISE because they do not like science, or do not know enough about science, or ISE.

As I have argued elsewhere, an assimilationist tendency informs a great deal of public science, springing from the belief that public science practices are inherently worthy and exclusion arises as the result of barriers to access (Dawson, 2014c). What I hope to have shown here is that social exclusion from public science is more complex, more intersectional, and more embedded than it might first appear and that only significant changes on the parts of practitioners, policy makers, and researchers can change core practices and patterns of exclusion. We must, therefore, develop forms of research and practice that seek to disrupt rather than reproduce these patterns of privilege and disadvantage in relation to science. If public science activities, ISE, or ISL are worth anything to our societies, then they must be inclusive.

REFERENCES

Ahmed, S. (2012). *On being included: Racism and diversity in institutional life.* Durham, NC: Duke University Press.

Becker, H. S. (1963). *Outsiders: Studies in the sociology of deviance.* New York: Free Press.

Bourdieu, P. (1998). *Practical reason.* Cambridge, UK: Polity Press.

Bourdieu, P., & Passeron, J.-C. (1990). *Reproduction in education, society and culture* (R. Nice, Trans., 2nd ed.). London, UK: Sage.

Brown, B. A., Henderson, J. B., Gray, S., Donovan, B., Sullivan, S., Patterson, A., & Waggstaff, W. (2015). From description to explanation: An empirical exploration of the African-American pipeline problem in STEM. *Journal of Research in Science Teaching, Advance online publication.*

Dawson, E. (2014a). Equity in informal science education: Developing an access and equity framework for science museums and science centres. *Studies in Science Education, 50*, 209–247.

Dawson, E. (2014b). "Not Designed for Us": How science museums and science centers socially exclude low-income, minority ethnic groups. *Science Education, 98*, 981–1008.

Dawson, E. (2014c). Reframing social exclusion from science communication: Moving away from "barriers" towards a more complex perspective. *Journal of Science Communication, 13*, 1–5.

Holland, D., Skinner, D., Lachiotte Jr, W., & Cain, C. (2001). *Identity and agency in cultural worlds.* Cambridge, MA: Harvard University Press.

Roth, W.-M. (2008). Bricolage, métissage, hybridity, heterogeneity, diaspora: Concepts for thinking science education in the 21st century. *Cultural Studies of Science Education, 3*, 891–916.

8 Teachers in the Outdoors
Bridging Formal and Informal Practices

Tali Tal, Mordechai Aviam, Rachel Levin Peled, Nirit Lavie Alon

The outdoor learning environment has many benefits for meaningful learning. Students can learn about real phenomena in natural conditions. They can freely explore, raise questions, and investigate. They can easily socialize and apply social learning. They can draw on personal experiences and use various funds of knowledge. However, teachers commonly refrain from taking their students out because of personal challenges, insufficient training, overloaded curriculum, testing routines, and school bureaucracy. Moreover, teachers often refrain from viewing scientific practices in a way that allows investigating the real world outside the classroom. In an attempt to help Israeli pre- and in-service teachers cope with these challenges, we purposefully went beyond the natural sciences and introduced teachers to different genres of investigation—archeological, social, and ecological—carried out in one area, while focusing on different dimensions of the system to understand its features and the relationship between them. We developed an online environment to support the teachers' enhanced collaboration and meaningful discussions.

At the core of this endeavor was the ambiguity that we identified in the way pre- and in-service teachers defined inquiry-based learning, scientific inquiry, or other forms of inquiry. Interestingly, this ambiguity became apparent in our analysis of the Israeli Ministry of Education's documents. Much confusion is reflected in these documents about "scientific research," "just research," or "study" and between quantitative and qualitative methods and the way they are presented to teachers and students (Levin Peled et al., 2015).

Our basic assumption was that by crossing the border between genres and disciplines, science teachers would be able teach inquiry in a more realistic and authentic way. We view inquiry learning as a method to teach content as well as discipline-specific reasoning, skills, and practices by engaging students in collaborative investigations that involve reading, generating questions, suggesting methods to address the question, collect and interpret data, draw conclusions and communicate them. We see inquiry learning as organized around relevant and meaningful problems. Students should be

cognitively engaged in sense-making, developing evidence-based explanations, and communicating their ideas. We aimed at using the outdoor environment to teach various practices and ways of investigating natural as well
as social and cultural phenomena. Instead of following the beaten path of
teaching "*the* scientific method" toolbox, we preferred focusing on inquiry
learning as a knowledge-building activity, which is informed by acknowledged theories in the field. We hope that by departing from the traditional
positivist model, which still dominates science education worldwide, the
teachers will deepen their understanding of how science happens in a real
context and will accept and tolerate a broader representation of scientific
inquiry in their classrooms.

WHY THE OUTDOORS?

First, all four of us have wide experience as outdoor educators working in
formal and informal education systems. We love the outdoors and believe
that much of what we have learned was in out-of-school environments. We
also believe that the school walls can be a metaphor for artificial boundaries
between learning and fun, between being active and passive and between
fields of knowledge or well-defined disciplines and interdisciplinary and
systems approaches. We believe that taking the teachers away from their
known teaching environment and comfort zone could help them find new
avenues for teaching, in general, and to inquiry-based teaching, in particular. Outdoor education has a long history and tradition worldwide. In Israel,
for many years it has had a special added value.

The history of out-of-school learning in Israel is as long as the history of
the Hebrew schools from the late 1890s. Unlike field trips in other countries, they had—and still have today—a strong national value, in addition
to the other more internationally common aims, in the cognitive, affective,
and social domains. In Israel, the field trip in the outdoors became a central means to creating a new nation. It was influenced by trends that were
developed in Germany in the 19th century as part of a romantic youth
movement ideology and by progressive radical educators who emigrated
from Russia and other Eastern European countries during the early 1900s.
The field trip to natural environments was viewed as a way of learning
about the natural history, but more so as a transition from (marginalized) European urban life to living in harmony with nature and becoming
attached to the land. The field trip became not only a central means for
teaching about the environment but also a means to develop a new set of
values that were influenced by the natural pedagogy, on the one hand, and
by emerging national thought, on the other hand. Apparently, these two
motives still shape the current policy of the Israeli Ministry of Education
regarding field trips.

WHAT IS INQUIRY LEARNING?

The first description of scientific inquiry in the *Inquiry and the National Science Education Standards* (NRC, 2000) is a story of Brian Atwater, a geologist who investigated the meaning of dead trees along the Pacific Coast in Washington. The description of his work opens a window on how inquiry is performed through observations, asking questions, looking for evidence from a variety of sources, conducting measurements, comparing with historical documents, suggesting explanations and validating them. The story emphasizes human curiosity, deep thinking, consistency, persistency, and communicating with an audience. This description is in line with characteristics of inquiry suggested by Roth and Bowen (1995), which we agree with: learning in context of ill-defined problems; experiencing uncertainties, ambiguities, and the social nature of scientific work and knowledge; and experiencing being part of a community of inquiry in which members can draw on the expertise of others, whether they are peers or teachers.

Previous scholarly work has already shown the multifaceted image of inquiry, inquiry learning, and inquiry teaching that involve epistemic and pedagogical aspects related to how inquiry is defined, what inquiry learning is, and how inquiry learning can be taught (e.g., Crawford, 2014). The scope of this chapter does not allow us to fully uphold the ideas related to inquiry-based learning and teaching; however, we do wish to address a few issues that are particularly relevant to our study. In defining what "scientific" is, with respect to learning through inquiry, most researchers in science education refer to the study of the material (or natural) world. We agree with the scholars who argue that this emphasis expresses the adoption of the physical science model and overgeneralizing it. Mayer and Kumano (1999) point out the "narrowness of the traditional discussions of the history and philosophy of science in the science education literature" (p. 77). This narrowness is reflected in the worldwide dominance of the physical science's images of inquiry in the science curricula. Such images are dated: "Grounded in narrow interpretations of positivistic views about scientific inquiry, the accepted scientific method is based on a view that doing science is doing experiments" (Grandy & Duschl, 2008, p. 307).

There now is a growing awareness that science itself has developed and much of what has been considered in the past to be perfectly reasonable and respectable science has limited relevance with respect to many of the socioscientific issues we face in the modern world. Already McGinn and Roth (1999) wrote that ethnographic studies on scientists' practices "support the claim that the 'scientific method' is largely a myth and does not describe what scientists actually do" (p. 15). They oppose the image presented in many books of "real scientists" and show how nonscientific factors influence research. A possible example for the gaps between traditional images of scientific inquiry and the aforementioned views can be an investigation of the relationship among lifestyles, public health, and specific diseases. Such investigations can be criticized as not scientific enough because of the elusive variable of "lifestyles" or because of how the data were generated and interpreted. However, we believe these issues are meaningful and relevant to students and to their lives as citizens.

To teach inquiry in more realistic and authentic ways, teachers should be given opportunities to confront and examine their beliefs about the nature of science in light of new paradigms and philosophies, they should experience reflective personal and group activities aimed at the development of alternative pedagogical approaches designed to address the issues arising from "post-normal" science, and they should adopt appropriate strategies for incorporating appreciation of complexity and systems thinking (Gray & Bryce, 2006). Our effort addresses these expectations.

In Israel, where our program took place, the education system is centralized under the Ministry of Education, which publishes the national curriculum, supervises the teachers, and is responsible for periodic reforms, national testing, and ongoing teacher professional development. In recent years, meaningful learning and learning through inquiry became central to the Ministry's work and discourse. There is much pressure on districts and schools to implement advanced pedagogies that encourage "meaningful learning," "inquiry-based learning," and other labels for student-centered learning. The program we developed is offered to secondary schoolteachers (grades 7–12) but a few advanced preservice students have enrolled in the program as well.

THE PROGRAM: GOALS AND DESIGN

Three inquiry activities carried out in the outdoors were the focus of this teacher-training program. The activities were in the fields of ecology (the study of the natural world and relationships with human impacts), sociology (the study of human societies), and archeology (the study of ancient cultures' material world). We carried out all the activities in one particular area to develop a more profound understanding of the relationship between the physical, the historical, and the social layers of a system. While standing at the top of Shikhin ruins, where the archeological investigation took place (see http://www.samford.edu/shikhin/), one could easily see the hill where the ecological investigation took place, one kilometer west, and the village of Hoshaya, where part of the sociological study took place, one kilometer east.

Our main goals were to (a) expand and deepen the teachers' views of inquiry and inquiry-based teaching and (b) encourage teachers to use the outdoor learning environment, not only to teach about things and/or to develop national values, but, moreover, to make them enjoy learning and teaching in the outdoors and to develop teaching for engagement and for understanding complexity. One more reason for learning in the outdoors, especially for science teachers, was to help them depart from the image of inquiry learning as experimentation. We used as well a variety of technological tools to enhance collaborative inquiry learning and reflection. Throughout the program, the teachers worked mainly in small groups, and sometimes individually, in the preparation stages of the investigations.

The entire program was designed in a way that every outdoor activity was preceded by preparation and was followed up by a wrap-up activity. All of the pre- and post-field trip activities took place individually or collaboratively on the website, as shown in Table 8.1. We view technology as a driving force that enables collaborative learning in various learning environments and especially in inquiry learning. Thus we designed a simple website to support the program, a website that allowed sharing and learning from each other, which could be built by teachers as well. The website was designed to support teachers in their efforts to teach through inquiry by providing a whole battery of scaffolds. We used available mobile apps for the fieldwork—ones that every teacher can download for free. Altogether, we

Table 8.1 The preparation and wrap-up activities.

	Preparation (online)	Wrap-up (online)
Ecology	Individual: Map analysis (geological, soil, vegetation, demographic, and archeological); reading background materials that we uploaded on ecological investigations' methods. Group: Recommending and uploading other resources; suggesting research questions. We presented a battery of measuring and data collection instruments and methods that the teachers could choose from for designing their group's study.	Group: Data analysis, generating a collaborative document, preparing a PPT presentation summarizing the investigation to communicate findings, and uploading it to the website. Individual: Choosing another presentation, critiquing, and offering suggestions (critique as part of inquiry); online reflection addressing the entire experience: preparation, fieldwork, teamwork, learning experience, generating further questions about inquiry and about the contribution of technology. Such reflections summed up all the investigations.

(Continued)

Table 8.1 (Continued)

	Preparation (online)	**Wrap-up (online)**
Archeology	Individual: Reading a real excavation report from an earlier season and learning about research methods: survey and digging and the kinds of evidence they provide. Group: Teachers were asked to go out and take photos of a few objects in the area that are small enough to put in a basket and create a PPT slideshow that tells what can be learned about our period (climate, lifestyle, culture) from the photographed objects. This was to model part of the archeologists' work. Create a class Knowledge Table of archeological periods to which each group contributed two periods.	Group: Creating a typological board (data analysis); summarizing the findings and making conclusions; creating a group video clip that presents the whole study (as another form of communicating science); individual reflection.
Sociology	Individual: We asked the teachers to look at social science research journals, choose one abstract from a research paper, and write the research goal/ questions, the methods used, the analysis, and the main findings. After this individual work, the groups were asked to suggest their research questions to be studied in one or both of the two villages. Having a science background, it was harder for them to come up with "social" research questions, even after reading about the villages and doing their preparation activity.	Group: Creating and uploading to the website either a scientific poster that presents the study or a conference abstract of 250 words.

worked with ten groups of three to four secondary-school science teachers during the 2013/14 school year. Most of the teachers were experienced and few were beginning teachers. The project is ongoing, but we are reporting here on its first year. In the next sections, we describe the three investigations the teachers were engaged in. Before the first investigation, the participating teachers were asked to form groups based on a collaborative Power-Point introduction presentation, to which each individual added one slide. In the next section, we elaborate on the fieldwork part of the three inquiry experiences.

The Ecological Investigation

The outdoor investigation was the first face-to-face meeting between the teachers and the researchers. The first half hour was dedicated to personal introductions and a short introductory explanation about the area. Then we all climbed up the hill of the archeological site for geographical orientation. On the way to the hilltop, we encouraged the teachers to point out "interesting phenomena" and ask questions about them. Some noticed a variety of insect pollinators on the wild flowers, some pointed out differences between the plant coverage on the hills around, one noticed a badger's den, and another asked why prickly pears grow on the hill. This was a good opportunity to bring up the idea that there was something else there and maybe another look at the ground would raise more questions. It did not take long for the teachers to realize that the ground was covered with small pieces of pottery and that this would be the site for the future archeological investigation. That part of the day ended with observing the ecological investigation site from a distance and identifying two different habitats that they could detect in the map analysis task as well. One side of the hill was covered by a low natural Mediterranean chaparral formation and densely planted pine trees covered the other side. The observation allowed for asking more questions and making hypotheses.

After having a picnic together and socializing, we went on to a stage in which each group could discuss and revise its inquiry question in light of what was seen on-site. A few groups revised their questions and others kept the ones they suggested online. We provided tape measures, pH-meters, LUX-meters, thermometers of various types, hygrometers, and field guides. The teachers were asked in advance to download various data collection apps to their smartphones, and we designed a data collection sheet in Google Forms to use in the field. We printed copies of the forms as well, in case working outside with the small screens would not be comfortable (which indeed it was not). Some topics (questions) investigated were the different vegetation in two soil types, the amount and diversity of annual vegetation on the two sides of the hill, and the distribution pattern of the local oregano and cyclamens with respect to rock coverage and amount of light. Although not studied by any of the groups, we pointed out another two areas in which human impact could be seen and studied: a hill that was once farmed and is gradually being taken over by natural vegetation, and a wide path in the natural chaparral caused by ATV riders who cause the run-off patterns to change and harm the natural vegetation. These observations allowed us to connect scientific data collection to man-made changes in the ecosystem and bring up questions about whether we study the natural world or the relationships between natural and social systems.

The Sociological Inquiry

The Galilee region, where our program took place, is geographically and demographically diverse. Jewish cities and villages and Arab cities and villages are built in close proximity to one another, creating unique religious and ethnic diversity within and between communities. In this investigation, we focused on two villages: an affluent, religious Jewish village and a large Bedouin-Muslim village. Most of the Bedouin villages in the country are characterized by their low socioeconomic status, even within the Arab population. The vast majority of the residents of the Jewish village had college degrees, whereas the vast majority of the Bedouin residents hardly had high school education. However, the community is proud of its growing interest in enrolling in higher education institutions and of having a growing number of college students in recent years.

At first, almost all the inquiry questions brought up by the groups were comparative: comparing the leisure-time activities in the two communities, comparing recycling dispositions, comparing education and employment patterns, and so forth. Only two groups out of ten suggested noncomparative designs: One group wanted to study the location-based tourism industry in the Jewish village, and the other group, who noticed the great number of cyclists in the area during the ecological investigation, wanted to investigate the "culture of mountain bikers" in the study area.

Only after further discussion in the fieldwork morning, and after meeting with representatives of the two communities, did some of the groups ask to revise their question to focus deeply on a particular phenomenon rather than comparing the villages. Eventually, about half of the groups focused on one village only, arguing that they could learn more deeply about the community and that a comparison between such different groups would be worthless and would not be interesting enough. Another transition was from the more classical social science study that looks at relationships between variables and uses survey data to an ethnographic study that requires the teachers to find people to talk with. Yet most of the groups did look at the relationships between variables, even when the sample was very small: for example, the relationship between marriage within the family in the Bedouin village and education, or the relationship between education and recycling habits. As indicated, one group chose to focus on the social phenomenon of "Friday mountain bikers" in the area. This group followed the bikers in a few spots and interviewed them and officials from the biking organization. At the end of the investigation, the group was able to tell the story of this unique, but diverse, group of people. Figures 8.1 to 8.4 show the teachers during their ecological and sociological investigations and the professional development website.

Figure 8.1   The cyclists investigation.

Figure 8.2   The tourism investigation.

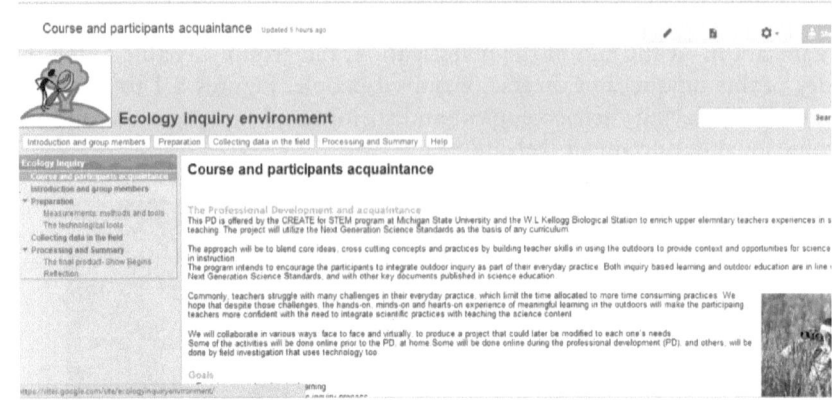

Figure 8.3   The professional development website.

Figure 8.4   The ecological investigation.

The Archeological Investigation

Unlike the previous investigation, for which the teachers could come up with a variety of questions, the given site that we were able to explore limited the questions that could be asked to ones such as: "Who were the inhabitants of the ancient settlement?" "What public buildings can be found?" "What evidence tells us the story of the settlement?" and "How did the ancient

Figure 8.5   Sifting.

Figure 8.6   Classifying.

Figure 8.7   Learning to dig.

Figure 8.8   Marking and labeling.

inhabitants make their living?" The investigation took place at the mostly Roman archeological site of Shikhin.[1]

In the field investigation, we worked in two larger groups, doing the survey and digging in turns. Much time was dedicated to teaching about and discussing the consistency of artifact (data) collection and to making sure the teachers' worked accordingly, accurately marking every specimen. This was important, as, in the discussions, many teachers expressed their surprise about how scientific, consistent, and transparent the work must be for artifacts to be regarded as evidence. They repeated these themes over and over while discovering the scientific aspects of the archeological research and in their references to the previous two investigations. In both the survey and the digging, we found many pieces of pottery, some of which were very interesting, which enhanced discussion in the whole group. Some of the work is presented in Figures 8.5 to 8.8.

BLURRING THE FORMAL AND INFORMAL DIVIDE

For us, the distinction between formal and informal aspects of education is difficult. Our position is supported by research that has found that the distinction is worthless, as some school-based activities can be regarded as informal, whereas often activities in museums are formal and inflexible (Hofstein & Rosenfeld, 1996). It is easier to argue that group picnics, sitting together in a circle, and sharing food are all informal or that a structured preparation activity, guided by instructions given in a website, is formal. However, most of the field activities are difficult to label. For example, a group of teachers who began their investigation in the Bedouin village were invited to a house, sat with the women, and listened to their stories. This group did not complete its pre-planned data collection but summarized the day as a great learning experience and, eventually, shifted its research to focus on the stories told by the Bedouin women. Another example is the pre-arranged meeting with the representatives of the villages. Although they were pre-planned, each of these meetings was then developed differently, according to the teachers' interests and the representatives' stories and emphases. The teachers then addressed the meetings:

> The meeting with Dr. Paz (in the Jewish village, Hoshaya) was an eye-opener. She told us facts that cannot be found in books or online. She told about herself and her family . . . since she's a faculty-member in Haifa University, she presented a comprehensive account. . . . The meeting with the head of the Education Department at the Bedouin village was important, too. I was impressed by the efforts of hosting us, providing food and showing the PPT. The presentation was less interesting than the opportunity to talk with him and listen to his stories about his childhood in a tent-home.

The next example of the blurring between formal and informal learning is from the archeological investigation, where the teachers met an archeological delegation from a U.S. university. A few teachers took the opportunity to talk with the American professor and his father (an archeologist, too) while their peers did the planned activity. This unplanned side activity ended with the teachers enthusiastically reporting to the others about what they had learned from the conversation. The last example is from the deep conversations that occurred at the end of each activity and continued online. In her reflection, M, who was the only one in her group with a social science background, reflected on the conversations in her group.

> The sociological inquiry brought me to [deal with] social sciences and qualitative methods, which made me happy. I like the "nosiness" with permission that characterizes this research. . . . I enjoyed less the debates in my group, especially because of disagreements about our methodology and the preparation of the questionnaires. All the others insisted on quantitative methods. This method is effective in some survey-based studies but lots of interesting data are lost. I love going out with already crystalized research questions and coming back with totally different ones after being out there. . . . The questionnaire didn't really reflect my positions and some questions were totally irrelevant. Eventually, after the field experience, and by agreement with all the group members, we integrated ethnography as well. This definitely empowered the inquiry, and us as researchers.

M's reflection taught us a lot about the images of inquiry held by science teachers. Given that almost all of the teachers' initial ideas for the sociology inquiry were comparisons based on survey data and variable analysis this was not surprising. All these examples demonstrate the kind of meaningful learning that does not require formal or informal labels. Space was given to the teachers to pursue their interests, guidelines directed their work, and assignments were designed to enhance deep learning that involves the epistemic, the social-dialogic, and the content aspects—all without clear reference to the extent of formality.

We collected much evidence of teachers acknowledging the importance of outdoor learning that promotes exploration, engagement, and curiosity—all of which are central to authentic science education that reflects a systems approach. Due to space limitations, Table 8.2 presents only a few of the teachers' quotes regarding their outdoor learning, classified by their emphases.

Table 8.2 Teachers' views of outdoor learning.

Contribution of outdoor learning	Teachers' quotes
Aesthetic experiences	The two things I will take with me from this experience (the ecology investigation) are: the Scarlet pimpernel (*Anagallis arvensis*) and the beautiful Oxythyrea noemi (beetle). The Scarlet pimpernel is one of my best-loved

(*Continued*)

Contribution of outdoor learning	Teachers' quotes
	flowers from childhood, whose name was unknown to me then. Only now, when someone put a guidebook in my hand did I look and find its name myself . . .
Exploring "terra incognita"	I liked the most finding and touching more than 2,000-year-old artifacts (part of a Hellenistic plate)—that I never thought I would do—just lying on the ground. I loved seeing how the way I see the space around me has changed (with some guiding from the experts, of course) when wearing "archeologist's lenses."
Active learning	I learned what fieldwork is all about. I have taken part in many investigations, but they were all in labs. I have participated in many field trips too, as a student, but I was always a passive observer and listener. There is no doubt, when you're an active learner outdoors, learning is more meaningful because you use your five senses to do the work and then you understand better—like I really understood the difference between the planted pine forest and the Mediterranean chaparral.
Curiosity and meaningful learning	As a teacher, I learned that even if I was listening to a lecture from the greatest expert in the field, I would not be as curious as I was and as thoughtful as I was while investigating in the field trip and in the follow-up analysis at home. In other words, I experienced very significant and meaningful learning out of school.
	I loved the meetings with the villagers (Bedouins). In the preparation, I did not believe I'd knock on someone's door and ask to speak with them, but the minute we arrived there and met the first family everything became clear to me . . .

CONCLUDING REFLECTION

The beautiful poem "The Boy of Winander" by William Wordsworth, glorifies the strong connections between youth and nature.

> There was a Boy; ye knew him well, ye cliffs
> And islands of Winander! many a time,
> At evening, when the earliest stars began
> To move along the edges of the hills,
> Rising or setting, would he stand alone,
> Beneath the trees, or by the glimmering lake;
> And there, with fingers interwoven, both hands
> Pressed closely palm to palm and to his mouth

Uplifted, he, as through an instrument,
Blew mimic hootings to the silent owls
That they might answer him; . . .

Pointing to the strong connection between youth and nature, which we might see as romantic nowadays, was not unique to scholars such as Rousseau and to poets and writers such as Wordsworth, Whitman, and Thoreau. This line of thought is clear in Dewey's writing in the early 1900s, and it is still present and strong in current writing in science education. Enjoyment in the outdoors was a central component in our effort, which was echoed in the teachers' reflections.

We believe that learning separate disciplines might undermine students' ability to connect ideas in science, and the NRC (2012) framework for K–12 science education deeply addressed this concern. Assuming that science education can benefit from looking at other modes of inquiry as well, we attempted to tie the three outdoor experiences into understanding one regional system by blurring boundaries between disciplines and between formal and informal education. The outdoor environment is not separated into disciplines, lesson schedule, and structured sequences. When standing in Shikhin and observing the area, social, historical, technological, and ecological concepts create one conglomerate and critical examination of all the variables that affected the region involves a variety of reasoning forms. After the experience our teachers went through, we still believe they need much support in investigating complex systems, but we hope that our outdoor inquiry program and investigation of real systems will not only help teachers teach outdoors, but will contribute to their classroom teaching as well. The students of these teachers are the real target population for this change, and we hope that they will experience meaningful learning in engaging environments that allow learning in ways that enable various forms of knowledge building.

ACKNOWLEDGMENTS

This research was supported by the Israel Science Foundation (grant No. 1010/13).

NOTE

1 For more detail see http://www.samford.edu/shikhin/.

REFERENCES

Crawford, B. A. (2014). From inquiry to scientific practices in the science classroom. In N. G. Lederman & S. K. Abell (Eds.), *Handbook of research on science education volume II* (pp. 515–541). New York, NY: Routledge.

Grandy, R., & Duschl, R. A. (2008). Consenzus: Expanding the scientific method and school science. In R. A. Duschl & R. E. Grandy (Eds.), *Teaching scientific inquiry* (pp. 304–325). Rotterdam, The Netherlands: Sense Publishers.

Hofstein, A., & Rosenfeld, S. (1996). Bridging the gap between formal and informal science learning. *Studies in Science Education, 28,* 87–112.

Mayer, V. J., & Kumano, Y. (1999). The role of system science in future school science curricula. *Studies in Science Education, 34,* 71–91.

McGinn, M. K., & Roth, W. (1999). Preparing students for competent scientific practice: Implications of recent research in science and technology studies. *Educational Researcher, 28*(3), 14–24.

National Research Council. (2000). *Inquiry and the national science education standards: A guide for teaching and learning.* Washington, DC: National Academies Press.

National Research Council. (2012). *A framework for K–12 science education: Practices, crosscutting concepts, and core ideas.* Washington, DC: National Academies Press.

Roth, W.-M., & Bowen, G. M. (1995). Knowing and interacting: A study of culture, practices, and resources in a grade 8 open-inquiry science classroom guided by a cognitive apprenticeship metaphor. *Cognition and Instruction, 13,* 73–128.

9 The Long-Term Influence on Three Classroom Teachers of Leading an After-School Science Enrichment Program
When Identities Converge

Phyllis Katz

The continual interaction of self and context has become known as identity development. Classroom science teachers learn to regard themselves and be regarded within the context of their schools. Schools are focused on providing evidence that the public trust of preparing citizens is fulfilled. The identity development of science teaching self-efficacy within the school context influences effective teaching and may benefit from work in an informal environment (McKinnon & Lamberts, 2013). What could motivate a career classroom teacher, unaware of the research, to build into her identity the work of leading a group of children in playful "informal" after-school science enrichment? In what ways did the after-school experience remain meaningful years later? What is to be learned about science teaching capacity as these teachers developed broadened identities?

This chapter presents three classroom teachers who chose to add periods of after-school science enrichment teaching to their workload 8–16 years ago. Two of the three are still teaching. The third and youngest is on maternity leave. For these women, expanding their participation in the science teaching community of practice had benefits for them as classroom teachers and suggests a continuum throughout their lives as they learn and teach about how the world works. For all of us, science learning continues through life as we each face decisions such as what to wear, what to eat, what treatments to seek, where to live, or how to vote. There are people who prepare educational experiences beyond the schools, and I am among them. We work in media, institutions, and community-based organizations. We must attract our audiences, unlike schools. We focus more on stimulating lifelong interest, preparing for multigenerational audiences who will pace themselves, providing environments schools cannot provide, taking time to play with the "what ifs" without any fear of failing a standardized test. Are there sufficient differences in these communities of practice to distinguish them as separate? The physically different settings would suggest that they are at least subcultures because of spatial designs or opportunities. The different systems of accountability would also suggest enough patterns of setting-specific behavior that the variations would yield distinctive identity development. How did these classroom teachers' identities merge the two teaching environments? Among the characteristics that the teachers share is an emotional devotion to lifelong learning—excitement—a goal absent from a major school science document.

I introduced the term, Continual Science Learning (CSL) to replace "informal" in my recent writing because we in the field struggle with "informal," as it suggests casual, unimportant, and not-school. None of these really address the pervasiveness and critical role science education plays in our lives (Katz, 2015). I therefore use CSL in this chapter to describe science teaching/learning beyond the school context.

CONTINUAL SCIENCE LEARNING CONTEXT

The CSL setting was a voluntary after-school science enrichment program I worked at developing, managing, and researching. It began locally and grew nationally in the US (with partial support from the National Science Foundation). Most of the adults who participated as Adult Leaders in the Hands On Science Outreach (HOSO) program from 1981–2007 were parents, graduate students, scientist volunteers, and others who did not primarily identify as science teachers. They valued science education and took our required preparation to use the guides and materials to provide small groups of elementary-aged children with weekly explorations and experiences in a relaxed environment emphasizing questions and playfulness. There were also a small number of classroom teachers who chose to lead groups. The three women in this study came from a large, well-regarded school system in the program's original site. These teachers have different cultural backgrounds, but commonalities in both their personal histories and practices became apparent as I sought insights to understand their choices to teach both in and out of school.

In exploring their science teaching identities, I interviewed the teachers about their personal histories and science teaching experiences. I analyzed documents that were provided by them and others about their teaching. In two cases, I observed their teaching to consider how their statements triangulated with their enactments. I have considered the long-term effects of teaching in the after-school program on these classroom teachers' identities by referencing the science education goal differences of two key (U.S.) National Research Council documents (Bell, Lewenstein, Shouse, & Feder, 2009; Duschl, Schweingruber, & Shouse, 2007). The teachers' stories speak to their dedication to model for their students the excitement of lifelong science learning. In the identity building that these teachers underwent, the CSL environment provided support for what these teachers found was missing in the formal education goals: a teacher and student emotional connection to learning.

THREE SCIENCE TEACHERS

Annie–The Excitement of Nurturing Long-Term Goals

Annie is a kindergarten teacher, now 50 years old. She loves being the first schoolteacher that her students will reference as they move through their school educations. She enjoys what the children can do. And she enjoys science, which she sees as an opportunity to teach while letting the children employ their natural curiosity. When she was nominated for the Presidential Award for Excellence in Mathematics and Science Teaching (PAEMST), her principal's endorsement letter said Annie "presents her love of learning in a fresh and energetic way. . . . Her kindergarten class is filled with happy, questioning, and alert faces." "Her students learn to think, ask questions, and to be satisfied only when all of their questions are answered. Whether in the playground," the letter continued, "or in the classroom, her students have a purpose and a mission with their teacher as their guide." The principal notes that she receives many letters of praise for Annie and quotes one. "I'm so impressed by the way she infuses the values of respect, kindness and appreciation into each day, often through the use of music. He [her son] has blossomed like 'the flower that shatters the stone. . . . What a wonderful and amazing teacher!'"

One of Annie's colleagues wrote in support of the same award that Annie led the discussion on school admission requirements for kindergarten and performed as recording secretary for the council—all as a volunteer. The support letter ends by saying that Annie "is a big picture thinker who understands and envisions long-term goals."

Annie described a classroom anecdote that tells of her pleasure in her work. One day, she agreed to include in her room the children of a colleague who had become ill and needed to leave her class:

> The visiting children could not believe that we were going to design and build something. . . . One very determined child from the other room was completely overwhelmed with the concept of taking his drawing of a tree and making it into a 3d model. I watched one of my students take him over to the scrap basket and suggest that he create a cylinder with a scrap of brown paper. This child could not believe I would REALLY let him try out his own ideas. You can only imagine the look on his face (and the smile in my heart) when he assured my student that he now understood.

Annie describes her history as that of a curious child who attended "an academically challenging, but very conservatively-based religious school." She says that she adapted by memorizing what she was expected to repeat but felt conflict with her own perceptions. Told that women could not work in her church, she knew that teaching was open to her.

She developed her own sense of how she would be as a teacher from negative experiences and, lacking role models, also developed a reticence. This lack of confidence followed her into her career years. Although she often referred to her earlier feelings of inadequacy, her actions displayed a willingness to do what she felt necessary when she confronted her ultimate goal of creating an environment where children felt safe and comfortable. They could then recognize their natural abilities to explore while she supported them in explaining. It was this determination to support her children's learning that surfaced in her decision to teach in the after-school program in 2000.

> I was working for a principal who had set up a suffocating environment for children. . . . An opportunity to improve the academic environment, work with children, encourage curiosity, keep learning on my own, and work around that administration. Count me in!

When Annie reflects on what she learned in the after-school program, she says that she found a place that told her she was not crazy to believe that children were "incredibly capable, perceptive and learn in unique ways, and that being nervous may be the same as being passionate with doubt added in" (Interview, January 3, 2015). The after-school science enrichment program required the collaboration of the schools (for space), the parents (for publicity and logistics), and the nonprofit business that was the program. Annie noted that children thrive when whole communities work together. She said that the after-school science program reinforced her teaching methods and provided her with support when she was challenged. When she was observed in the enrichment program, she was complimented for following the children's interests and not chastised for straying from the curriculum. "I had been tormented so much by different administrations," she said, "I was terrified of not having everything perfect." The supportive principal under whom she was then working let her use professional time to work as an after-school science program trainer when her talents were tapped.

After she worked with the after-school program for a few years, Annie remembers that she felt the confidence developing that she now displays. She would speak up on committees; she made an exemplary early childhood video lesson for her school system; she engaged her kindergarteners in a STEM conference, inspired by what she had learned young children could do in science during the after-school science program. She believes that her benefits from her CSL teaching were that she should

> Try. Speak up even when your ideas are different and you are nervous . . . share the joy of learning. No two classes will ever be the same, even if the content, materials and goals are the same . . . each class will walk down its own path. Enjoy each walk. . . . There are many opportunities I wouldn't

have even been open to if the after-school program hadn't helped me take the risks needed to keep sticking my neck out there for children.

She also noted that the after-school program office provided her with access to learning materials and "real mentoring as opposed to monitoring." The CSL environment was a safe place for her, without fear of career failure, to develop an identity as a happy enthusiast respecting children's natural curiosity.

Tanesha–Science Excitement and Cultural Contribution

Tanesha was born into an African American family about 30 years ago, the younger of two girls. She says that her mother was always curious about how things worked. Neither her mother nor she had much exposure to science as a discipline. Her mother taught her to help others. As a first-generation college student, Tanesha entered her university as an architecture major. She chose to fulfill a college requirement in the education department, where she found "a disconnect" between her fellow education class students and urban communities. She said that a diversity and equity voice was missing, and she felt strongly enough about it to change her major. She applied for the after-school science internship that was offered prior to her methods class to get more experience teaching. She states that science was not her thing, but knowing that she would have to teach science, she wanted to learn more. The internship took place in spring 2007.

Tanesha was interviewed as part of research to explore effects of the internship. She said that she learned the importance of asking open-ended questions, not answering the questions for the students, and allowing them thinking time to see what they came up with. But what impressed her most then was the attitude of the children:

> The thing that I remember most was getting ready for the students to come in . . . And I just remember being, not shocked, but utterly surprised at how excited they were to learn about science because I don't remember ever running to get to my science classroom. And so I take that from the experience, just their excitement. It gets me excited about it and kind of made me get more excited about being a part of it.

Tanesha noted that the internship encouraged her to want to teach science, because she did not feel as if she needed to lecture as she had envisioned. She said, "The kids already came in with information and there was a lot to work with," and continued, "We have a lot of very interesting conversations in class that just sparks from one student making a comment and another student starting a discussion off of what that student had to say." When I visited Tanesha in her classroom two years later, I observed a question board that had a pocket for index cards. Children could ask anything anonymously. She used a box in which they could put questions. She said that she reviewed the cards for trends and answers. It helped her monitor her own teaching, she said, as well as giving voice to shy students.

Tanesha also talked about the adult relationships within the CSL program. She stated that she "learned how to facilitate discussions with all of us in the room and I think it helped make it better." Tanesha said about her mentor:

> She already had experience with how she reached the students. . . . And she taught us a lot about wait time to propose a prompt and give them time to look at it . . . and a lot of her classroom management techniques, and just how she organized it. And when she didn't get through something it wasn't like the end of the world. . . . I always felt it [the time] was productive. I learned a lot about what to do and still make science interesting and fun for the kids.

In a three-year, follow-up interview, Tanesha reiterated the after-school program skills she had learned to use in terms of the classroom teaching in which she was engaged. She said, "The students do a lot better when it comes to understanding the material when they work in small groups and they're able to touch and feel what's going on." She continued, "When it's abstract ideas it's a little bit harder for them to grasp—and I found that when they can actually relate it . . . to something in their life, they have a better grasp on it."

Tanesha continued to talk about her goals. She said, "I want to make sure that I'm still excited about some science and can transfer that to the kids. I know my first year I was super, super excited and all the kids knew me as 'Oh, Ms. Robins she loves science . . .' especially to the girls, because a lot of them don't really think that science is for girls unfortunately. Yeah, and they have their idea of what a scientist looks like and so I like using that in my teaching and just being able to keep relatable, but still making it meaningful."

In terms of her identity, Tanesha saw herself as a teacher who needed to understand her content area. She noted that elementary teachers were generalists and that she wanted to know her specialty (science) well. Recognizing that she was not what she would call "an expert," she nonetheless aimed at increasing knowledge. She said, "I want to make sure that I'm able to guide the kids, or lead the kids into questioning and not telling them how to get to the answer. Because I think knowing how to get there and that whole process is important. I do think of myself as a learner of science. I think I'm always learning. I think of myself as a lifelong learner. It never stops."

Tanesha also provided information on how others viewed her. She said that in casual conversations with people they would remark, "Oh, you should be teaching science." In her large urban school, she quickly was asked to become a school science team leader, crediting her enthusiasm back to her internship. For Tanesha, the CSL context introduced to her identity a pleasure in science as a playful, yet important subject area. It also provided her with the experience she sought to build a variety of teaching skills.

Fran–The Excitement of Exploration and the Willingness to Risk

Fran, 51, was born to teenaged parents who did not have the opportunity to get a higher education. She describes herself as curious and outdoor-loving, inspired by her nearby grandfather. She enjoyed biology in high school and remembers that she always wanted to teach—she was one of those children who "played school" in her basement. Her parents discouraged her. Her mother had worked in a local school and wondered to her daughter, "Why would you subject yourself to that?" They thought Fran should aim toward something practical and well paying. Fran spent five years in the U.S. Air Force. She was able to study while in the armed service, but the majors were limited. She took an undergraduate degree in business management and worked in that field for a few years, later taking her master's. She enjoyed the money, but the desire to teach won out. She went back to school for a second master's in teaching. Her husband was supportive as she made the transition, although her parents were not (but now are). Fran is now an assistant principal who also teaches. She likes the blend of policy/management and student contact. For a few years, she worked at the school system's headquarters but missed the students.

In 2001, Fran talked about herself in a discussion for a book on community science:

> I am a lifelong learner—I think that you need that to be a good teacher—and I have a blast. Every year is full of surprises. I feel like I generate a lot of enthusiasm in my students. I hear that from them and from their parents. . . . I am curious, so we feed upon each other.
>
> (Katz, 2001, p. 104)

In an interview in December 2014, Fran talked about how she was drawn to the after-school science program 15 years ago when her daughter was five years old and her school needed adult leaders to run the program. She arranged to leave her school-time position a half hour earlier than usual to participate in the after-school science program. As a relatively new teacher, she learned that "open inquiry is not necessarily chaos," and that flexibility to "not know the answer," no testing, and smaller groups were advantages to this setting for teaching, some of which she could apply in her classroom. She said, "It was exciting to see light bulbs. The program was a complement to the professional development that the school system offered." And then she added, "I am a risk-taker." She articulated that formal science education is dictated from entities outside of the people experiencing it and that informal science education was more "joyful exploration," where the learner has more control. She then added, "the difference between the kids in my class deemed at risk and the kids who are deemed gifted and talented is the outside experience. The kids who are doing outside enrichment . . . have background knowledge and it is easier for them to make those connections. We need to get monies and pour it into parent education, society education."

Fran provided an example of her fifth-grade teaching. She asked her students if there was a correlation between a pumpkin's size and its number of seeds. She asked her students to design an experiment around the molding of the pumpkin as it aged, and she asked the children's parents to help the family decide when to throw away the pumpkin based on evidence. "As educational communities," she concluded, "we need many and varied approaches to students to engage them in science. We need the students to be curious kids at home and as teachers, to keep the sparks going. There is science all around you. Experience is not always in our classrooms. We set the stage for those experiences." Fran's identity as a "joyful explorer" was reinforced in the CSL context as she recognized and contributed to the enrichment that she feels makes a difference for students.

CONSIDERING IDENTITIES–THE CSL CONTRIBUTION

Looking at this data (the lives, practices, and expressions) from the perspective of differences between in- and out-of-school science teaching/learning presented in the National Research Council (NRC) reports, *Taking Science to School* (2007) and *Learning Science in Informal Environments* (2009), we can see that they differ by two strands in their statements, presented here in Table 9.1.

Table 9.1   Strand descriptions in the NRC books.

Taking Science to School	Learning Science in Informal Environments
1) Know, use, and interpret scientific explanations of the natural world	1) Experience excitement, interest, and motivation to learn about phenomena in the natural and physical world.
2) Generate and evaluate scientific evidence and explanations	2) Come to generate, understand, remember, and use concepts, explanations, arguments, models, and facts related to science.
3) Understand the nature and development of scientific knowledge	3) Manipulate, test, explore, predict, question, observe, and make sense of the natural and physical world.
4) Participate productively in scientific practices and discourse	4) Reflect on science as a way of knowing or processes, concepts, and institutions of science and on their own process of learning about phenomena.

(Continued)

Taking Science to School	Learning Science in Informal Environments
	5) Participate in scientific activities and learning practices with others using scientific language and tools.
	6) Think about themselves as science learners and develop an identity as someone who knows about, uses, and sometimes contributes to science.

What does not overlap in these learning goal strands from those who perceive science learning as different in each environment are strands "1" and "6" in the informal list. Those focusing on school learning list nothing about the excitement in science learning, nor the development of a personal identity as a participant, although this might be inferred from the successful accomplishment of the strands. What is the importance of the emotional qualities that make up strand 1 in the informal environments list?

Precisely because CSL is not compulsory schooling, CSL educators have always had to include exciting features that attract an audience and provide that audience with reasons to participate. This was true of the after-school program. Each of these three classroom teachers recognized that the continuity of their excitement in teaching science, reinforced by their after-school program participation, was a model for their students and a necessary part of their own identities as capable science teachers.

Identity in the science enterprise requires creativity that flows from emotions. "Historians, sociologists, and even biologists develop an understanding of the people or animals they study through empathy. Mathematicians and physical scientists achieve visual, muscular, and tactile intuitions by paying attention to the feelings that problems and pattern evoke," wrote Root-Bernstein (2000, A64). Emotions control our attention and decision making (Boyd, 2012). Thus emotions are essential to learning and teaching science despite the discipline's reputation for rationality.

What these three classroom teachers communicated about their self-selected after-school science teaching and learning experience was that it supported their preexisting identities as curious people who wanted to bring emotional connections to science to their students. The emotional content (or ISE Strand 1) that they included in their classroom teaching was in fact recognized and rewarded by others in each case, supporting their identities as successful science teachers. The teachers' different cultural backgrounds provide evidence that there are multiple paths in lifelong CSL that develop the resilient and persistent emotional connection to science education. Seeking the after-school experience extended their participation. The themes that emerged from studying their stories are that they:

- Were aware that they themselves were curious children, that this was positive, and that they wanted to encourage this in their students;

- Had challenges in their lives that resulted in resiliency that enabled them to find a way to overcome obstacles to participate in the after-school program;
- Wanted to teach for their students' learning and their own gratification;
- Found supportive mentor relationships and techniques in the after-school program;
- Sustained their learning/activities derived from the after-school program after 8–16 years;
- Were aware of the emotional content of their teaching (sparks, excitement, joyful exploration); and
- Melded their identities as formal and CSL science teachers to their satisfaction.

What CSL has to contribute is emotional stimulation/connection and identity development as a lifelong participant. These strands, absent in the NRC formal school report, are what these teachers bring to their students. These teachers have developed a common theory of affect in that they recognize in themselves an excitement in the curiosity that drives them and the need for their students to experience this. Each found in the after-school program permission and techniques to teach science in such a way that children would choose it, thus blending their school curricular teaching and CSL affective emphasis in their teaching identities. They found support for this theory in the CSL context designed for affective stimulation with its focus on enjoyable experimentation and creativity. The after-school science enrichment program therefore served as an opportunity to strengthen an identity component that was not supported in their school work.

POSSIBILITIES

The after-school program was a professional development opportunity. It gave these teachers additional experience in science teaching that was both playful and substantive. They took this complement to their school system professional development and wove it into their practices. The weaving that they created has survived for years as part of who they are because it was not an entirely new pattern but a strengthening of their science teaching identity from other experiences earlier in their lives, including the emotional connections to people who nurtured their curiosity and enthusiasm. From these three stories, there are possibilities. When we are recruiting science teachers, perhaps we should ask if they consider themselves curious and excited lifelong learners, as well as good students themselves. What examples of their own resilience and history can they provide? Do they see teaching as more than "following curriculum and raising scores"? Tanesha's experience suggests that including the two informal science education strands that are missing in the formal schooling document has implications for educational equity. Recognizing that modeling for and believing in her African American students' capacities to be curious science learners and to develop their own identities as science learners provides threads to the next generation from which they can weave their own successful stories.

These cases support existing research findings pointing to the significant role of informal science to science education and specifically to the development of science teaching identities (Avraamidou, 2014). They also contribute longitudinal evidence of the additive effects of CSL, which is often difficult to obtain (Dierking et al., 2003). I add my voice to those calling for redesign of teacher preparation and in-service programs to reflect the researched benefits of CSL. I would require CSL internships. These could be in institutions such as museums, zoos, aquaria, or botanical gardens. They could be science education media development. They could be community-based programs, such as the after-school program in this chapter. The experiences could ideally be community-based local improvements that depend on science knowledge, providing opportunities for science-in-society experiences. These internships would be given equal weight with practice teaching in classroom settings. We could require CSL experiences for promotion, recertification, and salary advancement. Just as our National Science Foundation asks for rotating program officers to oversee grant making because it brings recent experience of the field to the decision-making process for the advancement of knowledge, I would ask of the classroom teachers that they spend time seeing how science education occurs outside the school to develop identities excited about new perspectives for their teaching.

REFERENCES

Avraamidou, L. (2014). Developing a reform-minded science teaching identity: The role of informal science environments. *Journal of Science Teacher Education*, *25*, 823–843.

Bell, P., Lewenstein, B., Shouse, A., & Feder, M. (2009). *Learning science in informal environments: People, places, and pursuits*. Washington, DC: National Research Council.

Boyd, G. W. (2012). The body, its emotions, the self, and consciousness. *Perspectives in Biology and Medicine*, *55*(3), 362–377.

Dierking, L., Falk, J. H., Rennie, L., Andersen, D., & Ellenbogen, K. (2003). Policy statement of the informal science education ad hoc committee. *Journal of Research in Science Teaching*, *40*, 108–111.

Duschl, R. A., Schweingruber, H. A., & Shouse, A. W. (2007). *Taking science to school: Learning and teaching science in Grades K–8*. Washington, DC: National Research Council.

Katz, P. (Ed.). (2001). *Community connections for science education, history and theory you can use*. Arlington, VA: NSTA Press.

Katz, P. (in press). Formerly ISE: Preparation for continual science learning. In P. Patrick (Ed.), *Preparing informal science educators*. Dordrecht, The Netherlands: Springer.

McKinnon, M., & Lamberts, R. (2013). Influencing science teaching self-efficacy beliefs of primary school teachers: A longitudinal case study. *International Journal of Science Education, Part B: Communication and Public Engagement, 4,* 172–194.

Root-Bernstein, R. (2000). Learning to think with emotions. *Chronicle of Higher Education*, January 14, 2000; 46, 19, Research Library Core, A64.

10 Integrating Mobile Technologies Into Outdoor Education to Mediate Learners' Engagement With Nature

Lucy McClain & Heather Toomey Zimmerman

Within both formal and informal settings, educators are increasingly integrating mobile technologies into outdoor learning opportunities to mediate people's experiences with the natural world. Research has demonstrated the positive effects of mobile technologies in outdoor spaces for supporting scientific inquiry with school students (e.g., Chen, Kao, & Sheu, 2003), yet less focus has been spent demonstrating how mobile technologies can support learners in their out-of-school time. Consequently, this work leverages both school- and non-school-based studies on technology-enhanced learning to advance research on out-of-school, technology-enhanced learning. In this chapter, we present recommendations from research on how to integrate mobile computers into outdoor learning opportunities. We developed four design guidelines for mobile computer integration into outdoor informal programs to support science learning and engagement:

1. Place-based observational questions;
2. Place-based textual prompts for focusing observations;
3. Drawing activities to record observations; and
4. Place-based images used to identify biota in the outdoors.

To illustrate these recommendations, we present findings from our larger study of 191 people using a place-based, technology-enhanced learning tool, an electronic trail guide ("e-trail guide"). We selected one family of four individuals as a focus for this chapter. Through this case study of one family's engagement patterns with technology and nature, we show how technologies can support learners to observe and learn about local plants and animals.

TECHNOLOGY-ENHANCED LEARNING IN INFORMAL EDUCATION SPACES

We approach this work with a sociocultural view of learning. Sociocultural learning theories assert that people make sense of their world with the people and cultural tools within the space where learning takes place. In this regard, learning is not a "predetermined system of activity but is a process that is constantly shaped and influenced by external factors such as one's cultural environment and the use of tools to achieve an objective" (Vygotsky, 1978, p. 55). As such, understanding learning requires researchers to examine previous learning experiences, norms, practices, and interests that help influence informal learning activities alongside a consideration of all aspects of the learning environment.

Learning and Engagement in Informal Spaces

Informal learning research often considers learners' *engagement* as a proxy for science learning. In our work, we build from the notion of engagement that includes people's personal involvement in, orientation toward, and socio-historic participation of a learning event. Additionally, we look to Heath and vom Lehn's (2008) definition of engagement in out-of-school spaces as the extent to which people engage in "interactivity" with others and with exhibits. Interactivity refers to both tactile investigations and social communication with others and, as the authors argue, is a critical component of engagement. Engagement, then, we define as a blend of people's orientation, participation, and interactivity toward the outdoors as mediated by handheld technologies.

Assessing for Engagement With Nature via Mobile Devices

In this study, we focus on one cultural tool, a mobile device, to understand how technology can mediate learning about and engaging with the natural world. An advantage of mobile-based learning is that an educational activity can transition between settings, such as indoor to outdoor, so that the learner is not confined to only one, permanent location. Consequently, mobile devices can enable new learning experiences that are grounded in authentic settings, such as the outdoors. Whereas the advantages of mobile-based learning are many, there is also concern that when mobile technologies are introduced, the device itself can distract learners. In light of these concerns, our e-trail guide was designed with the goal of engaging learners with nature in such a way that the mobile device *facilitated* observations of the physical environment, rather than creating a distraction that required learners to fixate on the screen.

Based on the literature, we have identified three types of technology-enabled learning engagement that we use within our work (McClain, 2015): (a) observation, (b) pointing, and (c) tactile investigation. The literature surrounding these patterns of engagement influenced our design choices aimed at fostering learners' engagement with the natural world. In this chapter, we focus on two of these nature-technology-social engagement patterns, observation and pointing, to assess the effectiveness of the design guidelines within our e-trail guide that include place-based observational questions, place-based textual prompts for focusing observation, drawing activities to record observations, and place-based images used to identify biota in the outdoors.

Observation. Technology-enabled engagement as observation occurs when learners view objects in their surroundings more often than they gaze at the screen. Researchers have recognized the importance of designing mobile programs to promote "heads-up" behaviors where learners are engaged not with a screen but with making observations of their surroundings (Eliasson, Pargman, Nouri, Spikol, & Ramberg, 2011). The "heads-down phenomenon" (Hsi, 2002), which has often been observed in museum spaces where visitors' attention is not on their peers or the site but on the mobile computer or exhibit screen, is a design challenge for informal educators wishing to implement mobile programs into their educational space.

Pointing. The second form of engagement, pointing, is a social sign of interaction with nature. Pointing, a form of gestural deixis, is a physical movement that combines with speech to orient others toward an object of interest (Roth, 2001). In education, pointing can play an important role in collaborative learning activities, while it also serves as a mode for externalizing one's thinking and understanding of a concept. In the space of a museum, one may observe a variety of museum visitors (e.g., school groups, families, individuals) while interacting with computer-based exhibits and conclude that certain gestures, such as pointing, were important modes of communication and sense-making in informal education spaces.

Technology Design Features Enabling Place-Based Learning and Engagement

The design of our e-trail guide was heavily influenced by place-based learning theory, which considers the importance of connecting learners with their community by anchoring pedagogy within the context of the local natural and social ecosystems (Gruenewald & Smith, 2014). Place-based education in biology can provide important learning opportunities and explorations of the local ecological landscape that can later serve as the foundation on which investigations of more distant or abstract phenomena can be constructed. To support a place-based learning experience, we included the following four design features to help orient and focus observations while on the nature trail:

1. Place-based observational questions;
2. Place-based textual prompts for focusing observation;
3. Drawing activities to record observations; and
4. Place-based images used to identify biota in the outdoors.

Lastly, the enactment of a self-guiding mobile device to support family and peer-peer learning and engagement with the outdoors was an important design principle in the development of a place-based e-trail guide. Previous literature reviews have made technology design suggestions for promoting parent-child learning interactions: posing questions and structuring joint tasks. Our research project elucidates how these forms of engagement might influence the use of e-trail guide to learn about trees together in their out-of-school time.

A PLACE-BASED TOOL FOR INFORMAL LEARNING: SHAVER'S CREEK E-TRAIL GUIDE

As part of our larger research and development agenda, the first author created a place-based tool for informal learning that was specific to a nature center associated with Penn State University, the Shaver's Creek Environmental Center (SCEC), which is located in the mid-Atlantic region of the US. SCEC receives thousands of visitors annually. Many kilometers of hiking trails begin on SCEC's property and wind through state-owned forestland, including the Boardwalk Trail. This case study focused on one family's learning experience about trees at one area along the Boardwalk Trail (Figure 10.1).

SCEC's e-trail guide was created using iBooks Author for iPads. The e-trail guide curated one nature trail: a 0.5 km/0.23-mile part of the Boardwalk Trail.

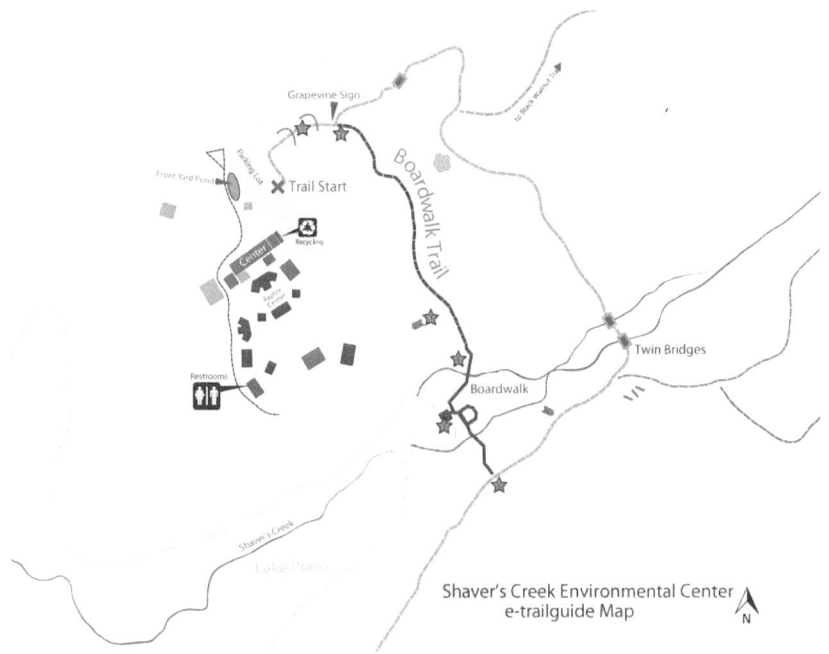

Figure 10.1 Shaver's Creek Environmental Center property map featuring Boardwalk Trail.

The e-trail guide provided content for six discovery spots along this trail; each discovery spot was presented as a book chapter in the e-trail guide. Seasonal information and digital content such as videos, photos, and audio-recordings were included.

The e-trail guide was designed to facilitate learners' engagement with the natural world that included a coordination of the learners' senses, the science content of the e-trail guide, and the learners' surroundings. The iBook software provided a means to include interactive features, called widgets. Widgets for drawing, entering data, and recording field notes were included.

In the e-trail guide, each discovery spot and on the move content section includes a variety of activities to engage the learners: questions, prompts, widgets, challenges, quizzes, and joint activities. Because one discovery spot (#3, Juniper's Bench) included the most activity variety of any of the other discovery spots (described next), we present this location to elucidate which components of the e-trail guide enabled certain patterns of interaction and dialogue for children and families.

In the e-trail guide, Juniper's Bench included information about deciduous and evergreen trees, questions and prompts related to observation of trees, a sketchpad widget for drawing a tree of the family's choice, place-specific tree identification images, and a photo widget. Upon first arriving to Juniper's Bench, the text in the e-trail guide asks, "What do you hear? What do you see?" These questions encouraged children and families to familiarize themselves with this area of the nature trail (Figure 10.2).

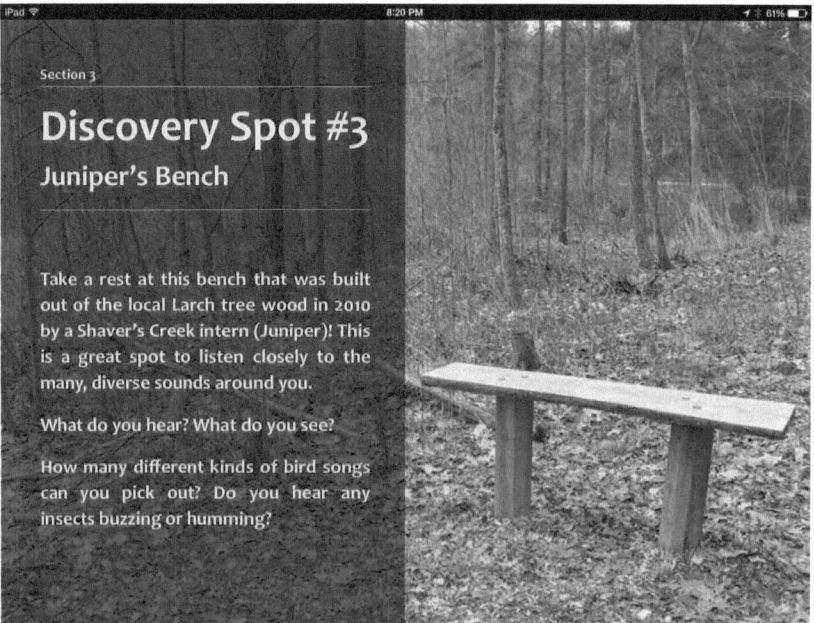

Figure 10.2   Introductory page to Juniper's Bench in e-trail guide.

Next, the e-trail guide provided an overview on deciduous trees and evergreen trees (Figure 10.3). To encourage observation, both a place-based question and a place-based prompt were included on this page. To focus observation on deciduous trees and their leaves, the e-trail guide asks, "Do you see any deciduous leaves on the ground around you?" Then, to shift their observation to the evergreen trees, the e-trail guide prompts, "To see the evergreen trees in this area, look closely beyond the leafy, deciduous trees!"

On the following page (Figure 10.4), children and families are asked a series of place-based questions meant to hone their observation skills of the native trees in that area. For the drawing activity, the text asked groups to work together to find a tree that interested them, draw the tree, and use the information in the e-trail guide to decide if their chosen tree was deciduous or evergreen.

After they completed the drawing, the next page showed and identified six species of tree leaves found in the area at this spot along the trail. The families and children were then asked to use this information to identify the tree they had previously drawn. If they could not identify their tree, the group members could take a photo of their tree (Figure 10.5).

DECIDUOUS TREES VS. EVERGREEN TREES

Deciduous trees have big leaves that fall off the tree every fall and then come back in the spring! Do you see any deciduous leaves on the ground around you?

Redbud tree leaves

Evergreen trees stay green all year-round. Instead of leaves, they have needles that stay on the branches all year.

To see the evergreen trees in this area, look closely beyond the leafy, deciduous trees!

Blue Spruce needles

Figure 10.3   Information on deciduous trees and evergreen trees at Juniper's Bench in e-trail guide.

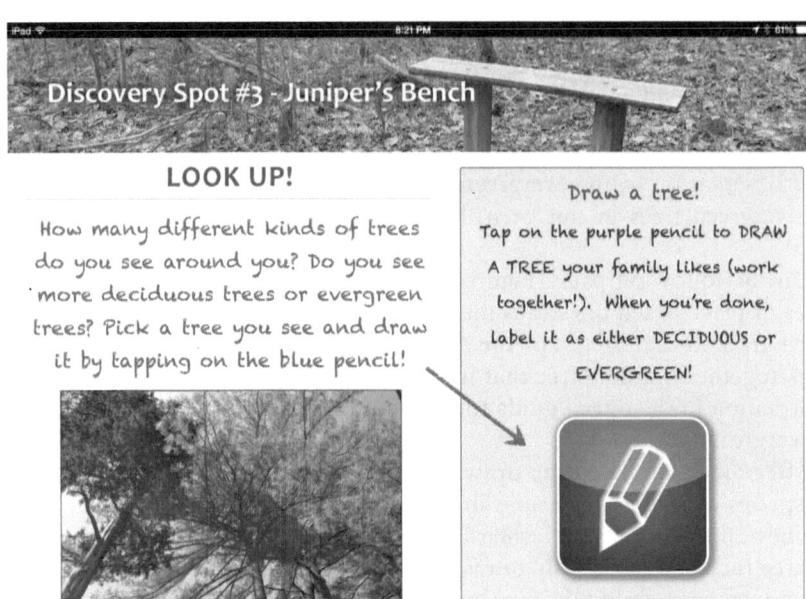

Figure 10.4 Drawing activity at Juniper's Bench in e-trail guide.

Figure 10.5 Place-based leaf images in e-trail guide used for tree identification at Juniper's Bench.

CASE STUDY OF THE DESIGN GUIDELINES TIED TO ENGAGEMENT

To further illustrate engagement in the outdoors, we present data from a study that leveraged a qualitative-based research method to investigate families' interactions with and conversations about trees during their walk on a nature trail while using an e-trail guide. The larger study included 32 families (*N* = 108 individuals) and small groups of children in summer camp (*N* = 83 individuals); we focus on one family here to illustrate perspectives important to others interested in supporting learning with technology in out-of-school time—especially the outdoors.

Focal Family

The Rodericks were strategically sampled for this case study because of their equanimity with each other and their equal involvement in the activities along the trail. In their pre-study online survey, this family indicated that they spend a lot of time outside; the daughters spend time outside on a daily basis; and as a family, they spend time outside two to three times per week taking short hikes, biking, and playing sports. Additionally, this family responded that they owned eight different technological devices, including an iPad and smartphones, yet did not have experience using these technologies in an outdoor setting. As such, using the e-trail guide on the nature trail at SCEC was a novel experience for this family. The Rodericks were (shown in Figure 10.6):

- Mom–worked as a human services director at the time of this study;
- Dad–worked as an electrician at the time of this study;

Figure 10.6   Roderick family at Juniper's Bench.

- Kendra–was 12 years old at the time of the study and entering seventh grade at a private school (wearing a white T-shirt as shown in Figure 10.6); and
- Whitney–was 9 at the time of the study and entering fourth grade at a public school (wearing a black T-shirt as shown in Figure 10.6).

Case Study Findings

Findings of this study elucidate four design features within the e-trail guide that facilitated engagement between the learners and the natural world. Previous studies have provided important guidelines for the heads-up engagement motivated design of the e-trail guide (Eliasson et al., 2011) and through our work, we add recommendations aimed at engaging people with nature as evidenced through observing and pointing to local flora and fauna. In the case study with the Roderick family, we saw numerous instances of engagement with nature through observation and pointing. This was also consistent across our whole data set. To illustrate our findings, short transcriptions from the data set are included in the following sections. Learners' gestures are represented through double parentheses [(())]. Brackets indicate overlapping talk between individuals ([). **Bold font** represents the application of our interpretation of the family's technology-mediated engagement with nature.

Place-based observational questions. Previous museum-based research studies have explored the role of science inquiry-based questions in supporting parent and child learning. According to Lyons, Becker, and Roberts (2010), "The role of technology in supporting science inquiry and expanding question-posing strategies in informal parent-child interactions has not yet been fully explored" (p. 94). In our work we sought to explore this topic in terms of how place-based questions created sense-making opportunities for family groups on a nature trail. We found that questions pertaining specifically to the area in which the Rodericks were located along the trail facilitated their observations of the natural biota immediately surrounding them, as exemplified in the following vignette:

> *KENDRA:* ((Reading from the e-trail guide)) Deciduous trees versus evergreen trees. Deciduous trees have big leaves that fall off the tree every fall, then come back in the spring. Do you see any deciduous leaves on the ground around you? [PLACE-BASED QUESTION]
>
> *WHITNEY:* What's a dec-
>
> *KENDRA:* Yes, there's one. ((**looks down** to ground and **points** to ground))
>
> *MOM:* ((Laughs)) Yes, you're right. There's one. ((**points** with toe at leaf on ground))
>
> *WHITNEY:* There's one, there's one, there's one. ((**points** at several leaves on ground))

The simple question, "Do you see any deciduous leaves on the ground around you?" provoked each of the family members to look at the ground, and Kendra first points out a leaf nearby, "Yes, there's one" and her mom affirmed her observation then pointed out a leaf next to her by pointing at it with the toe of her shoe. Whitney followed suit and repeatedly pointed at three different leaves nearby on the ground while saying, "There's one, there's one, there's one." In this example, the acts of both actively looking at the ground for deciduous leaves and pointing in response to the question stated in the e-trail guide signified the family's connection between the mobile-based content to the physical environment of the nature trail.

Place-based textual prompts for focusing observations. With the e-trail guide, our prompts were designed to focus users' attentions on certain features of the landscape. To illustrate one example of how interactions with nature were mediated through the technology of the e-trail guide, the following excerpt captures the family observing the trees around them after Kendra reads a place-based prompt out loud upon their arrival at Juniper's Bench:

> *KENDRA*: ((Reading from the e-trail guide)) Evergreen trees stay green all year-round. Instead of leaves they have needles that stay on the branches all year. To see the evergreen trees in this area, look closely beyond the leafy deciduous trees. [PLACE-BASED PROMPT]
>
> ((Each family member **looks up**))
>
> *DAD*: What kind of tree was it again?
> *MOM*: Like a pine tree or hemlock tree.
> *KENDRA*: ((turns around and **points**)) Is that . . . ? No that's not it.
> *WHITNEY*: Is that a pine tree? ((**points**))
> *DAD*: ((**Points**)) I found one. ((Everyone turns to see where he is **pointing**))
> *MOM*: Where? ((**leans in to see** where Dad is **pointing**))
> *DAD*: Right back there ((**still pointing**))
> *KENDRA*: Oh yeah! I see it!
> *MOM*: Oh yeah! ((**points** at the tree)) Good job, Dad! Yeah, I didn't see that at first.

After the Rodericks arrived at Juniper's Bench, they were prompted to observe the different types of trees around them. While Kendra read out loud from the e-trail guide about evergreen trees, the rest of the family listened and examined the setting, presumably to connect the e-trail guide content to their observations. When Kendra read the place-based prompt, "To see the evergreen trees in this area, look closely beyond the leafy deciduous trees," each family member looked up to find an evergreen tree. Dad asked others for help to find an evergreen tree; Mom restated the prompt, "Like a pine tree or a hemlock tree," which are two specific kinds of evergreen trees. Dad was the first family member to find an evergreen tree; he pointed at the tree and said, "I found one." When Mom asked, "Where?", she leaned in closer to see where Dad was pointing and Dad responded with "Right there" as he maintained his pointing gesture for the others to follow with their eyes. Kendra confirmed her Dad's find when she says, "Oh yeah! I see it!" and Mom pointed toward the tree as she excitedly says, "Oh yeah! Good job, Dad!" In this episode, each family member engaged with sustained heads-up observations and at least one instance of pointing as they sought out evergreen trees together based on the e-trail guide's prompt.

Drawing activities to record observations. Observation is deemed an important and foundational skill for all practices of science, but it can be difficult for young children and novices to observe the world in the way that scientists do. To support observational practices, we build on the idea that drawing is influential for honing young children's observation skills. In their work with 42 kindergarten students observing live animals, Fox and Lee (2013) reported that the children who drew their observations (versus those who did not draw) scored higher on descriptions of the animals. Because drawing can facilitate closer inspections of one's surroundings, we included multiple opportunities to draw on a sketchpad widget included in the e-trail guide. At Juniper's Bench, the main activity is using the sketchpad widget to draw a nearby tree that the family chooses. When the Rodericks engaged with this activity, Whitney asked to be the digital artist and elicited the help of her family, as illustrated in the following section:

> *WHITNEY:* So what are we doing?
> *MOM:* Draw a tree.
> *KENDRA:* Draw a picture of a tree you see. [PLACE-BASED ACTIVITY]
> *DAD:* You have to know what you're drawing, though. Name it.
> *KENDRA:* And then label it deciduous or evergreen.
> *MOM:* Unless you draw a pine tree, they're all deciduous.
> *WHITNEY:* What if I drew one of . . . like that tree? ((**points** up, Kendra and Mom **look up** to where she is pointing))
> *MOM:* Yep! That's a deciduous.

Here Whitney asked for clarification for this activity, and Kendra and her mom restated that she was to draw a tree. Mom reminded her that, "Unless you draw a pine tree, they're all deciduous." Before she began her drawing, Whitney asked "What if I drew one of . . . like that tree?" and pointed up to a tree. Kendra and Mom looked up to where she was pointing and Mom confirmed that she had selected a deciduous tree to draw. The opportunity to select a tree and document it through drawing encouraged the Rodericks' engagement with nature through observation of the surrounding trees, discussion of the different types of trees in the area, and gesturing toward a specific tree of interest.

Place-based images used to identify biota in the outdoors. Our research team's previous work with family groups on the nature trails at Shaver's

Creek with field guides and laminated Pocket Naturalist© guides illumi-nated the frustration that can stem from being unable to identify plants, insects, or animals on the trail while using the cartoon drawings shown in many paper-based field guides (Zimmerman & McClain, 2014). As a result, the photographs included in the e-trail guide were all collected locally to the environmental center where this study took place. To show the utility of this, an excerpt is shared from the Roderick's time at Juniper's Bench. The Rodericks have completed a drawing of a tree at Juniper's Bench, and they are about to identify that tree by referencing images of six different tree leaves shown in the e-trail guide:

> *KENDRA*: ((Reading)) Now that you've drawn your tree, can you identify what kind it is by looking at the leaves?
> *MOM*: Oh my. ((**looks up** at trees))
> *WHITNEY*: Ooh, let's see!
> ((Mom sits down on bench next to Whitney to look at images of tree leaves in e-trail guide.)) [PLACE-BASED IMAGES]
> *MOM*: I think it's American ba—
> *WHITNEY*: I don't know . . . it's not, I don't think it's American basswood.
> *MOM*: Oh, I think it's American basswood. ((**points** up at tree in front of them))
> *KENDRA*: I think the one behind us is American basswood. ((turns body on bench and **points** behind them))
> *MOM*: ((**Looks** to wear Kendra pointed)) Oh yeah, I think you're right.

This episode was rich with engagement moments that are mediated through the e-trail guide. When Kendra read the instructions for the tree identification challenge, each family member moved closer to Whitney, who was holding the iPad, to get a better look at the photographs of the tree leaves. The fam-ily aimed to determine if the tree that Whitney drew was an American bass-wood. Whitney did not think the tree was an American basswood, but her mother did. As a point of comparison, Kendra pointed to and suggested that the tree behind them was an American basswood, to which her Mom agreed. While the family never explicitly came to a final conclusion as to what tree Whitney drew, they were able to use the place-based images from the e-trail guide to observe and identify the native trees surrounding Juniper's Bench.

In each of these data excerpts, specific designs within the e-trail guide were catalysts for the family's engagement moments. Prior research focus-ing on how to facilitate place-based learning in outdoor environments using mobile computers has recommended organizing the mobile content to con-nect with the local ecosystem, using images to amplify learners' observations of biota within their community, and providing opportunities to develop artifacts of local species through photographs or other means of data col-lection (Zimmerman & Land, 2014). Building from this work, our e-trail guide provides detailed design recommendations that promote engagement with nature. Simple place-based questions and prompts, a drawing activity, and place-based images used for identification were all effective methods for enabling the family members to observe their surroundings and collaborate on their observations through pointing out natural objects to one another.

APPLYING OUR RESEARCH TO PRACTICE

Throughout our work, we offer design guidelines that can offer a starting point for nature centers, museums, and others who want to engage youth and families in outdoor science learning. Our principles have been explored in two settings: (a) summer camp and (b) recreational weekends and evening visits or summer vacation-type visits by families. We opted to use an e-book format because our nature center partner did not have reliable cellular reception or wireless Internet access on its trail; however, our work provides guidance with regards to how environmental centers, natural areas, and parks can design mobile programs more broadly so as to engage visitors with nature during informal education experiences. To this end, we developed four design guidelines, which are featured in Table 10.1, to show our design intention, the example in our e-trail guide, and the learning outcome observed.

Table 10.1   Four design guidelines to engage visitors with nature during informal outdoor education experiences.

Design guideline	Example in our work	Engagement and learning outcomes observed in our work
(1) Place-based observational questions	Written text in the e-trail guide that asks, "Do you see any deciduous leaves on the ground around you?"	• Individual observations followed by pointing as a means of communication and understanding of the e-trail guide's content to other family members. • Joint attention and conversation around scientifically relevant objects.
(2) Place-based textual prompts for focusing observations	Written text in the e-trail guide that states, "To see the evergreen trees in this area, look closely beyond the leafy deciduous trees."	• Individual observations followed by pointing as a means of communication and achievement of the e-trail guide's task to other family members. • Joint attention and conversation around scientifically relevant objects.

(Continued)

Design guideline	Example in our work	Engagement and learning outcomes observed in our work
(3) Drawing activities to record observations	Sketchpad widget in the e-trail guide used for drawing a picture of a tree nearby.	• Shared observations and conversations about local tree species followed by pointing as a means of communication and demonstration of understanding of tree species. • Joint attention and collaboration in the task at hand by family members.
(4) Place-based images used to identify biota in the outdoors	Locally collected photographs of tree leaves shown in e-trail guide that assisted in tree identification.	• Shared observations and discussion of local trees as mediated by images in e-trail guide followed by pointing as a means of communication and orientation toward specific tree types to family members. • Joint inquiry and sense-making around scientifically relevant objects.

In this chapter, we present one family using a self-guiding e-trail guide because this family was typical of the others in our data set who used mobile technology as a facilitator of science learning and engagement with the outdoors. By focusing on one family using the e-trail guide at one discovery spot along the nature trail, this study was able to take a deep analytical dive into the data to determine not only what interactions between the family members and nature were mediated by the technology but also what specific features of the technology promoted these interactions. Four features of the e-trail guide emerged as being effective strategies for engaging family groups, like the Rodericks, with their natural surroundings at Juniper's Bench: place-based observational questions and place-based textual prompts for focusing observations facilitated the Rodericks to further observe their surroundings along the nature trail, whereas a drawing activity to record observations and place-based images used to identify biota in the outdoors furthered their engagement *and* collaboration about what they were observing at that spot along the trail.

CONCLUDING RECOMMENDATIONS FOR INFORMAL TECHNOLOGICALLY ENHANCED PEDAGOGY

For environmental education centers that wish to engage their visitors with nature in a meaningful way, this study found that providing an e-trail guide to families and children in summer camps was an effective means of fostering learners' engagement in the outdoors. To support science engagement, we recommend the inclusion of purposeful and place-based questions and textual prompts to focus observations on specific features of the landscape, drawing activities for recording observations, and place-based images to be used for identification purposes.

A challenge for environmental educators is introducing mobile-based learning in such a way that it enhances people's experiences outdoors, rather than creating a dependency on the device's screen. This case study, which focused on the implementation of a mobile-learning design set within a forested trail, elucidated the patterns of engagement between one set of family members and the natural world when a mobile-learning tool in the form of an electronic book was introduced. Results from this study provide important design considerations meant to promote different kinds of learner engagement with nature through emphasizing the importance of place-based pedagogy.

REFERENCES

Chen, Y. S., Kao, T. C., & Sheu, J. P. (2003). A mobile learning system for scaffolding bird watching learning. *Journal of Computer Assisted Learning, 19*, 347–359.

Eliasson, J., Pargman, T. C., Nouri, J., Spikol, D., & Ramberg, R. (2011). Mobile devices as support rather than distraction for mobile learners: Evaluating guidelines for design. *International Journal of Mobile and Blended Learning, 3*, 1–15.

Fox, J. E., & Lee, J. (2013). When children draw vs when children don't: Exploring the effects of observational drawing in science. *Creative Education, 4*, 11–14.

Gruenewald, D. A., & Smith, G.A. (2014). Introduction: Making room for the local. In D. A. Gruenewald & G. A. Smith (Eds.), *Place-based education in the global age: Local diversity* (pp. xii–xxiii). New York, NY: Taylor & Francis.

Heath, C., & vom Lehn, D. (2008). Configuring "interactivity": Enhancing engagement in science centres and museums. *Social Studies of Science, 38*, 63–91.

Hsi, S. (2002). The electronic guidebook: A study of user experiences using mobile web content in a museum setting. In *Proceedings of the IEEE International Workshop on Wireless and Mobile Technologies in Education* (pp. 48–54). Vaxjo, Sweden: IEEE Computer Society.

Lyons, L., Becker, D., & Roberts, J. A. (2010). Analyzing the affordances of mobile technologies for informal science learning. *Museums & Social Issues, 5*, 87–102.

McClain, L. R. (2015, April). Facilitating outdoor family learning on a nature trail with an e-trail guide. In H. T. Zimmerman (Chair), Structured poster session: Design principles for supporting family learning: Implications from nine empirical studies. *Annual Meeting of the American Educational Research Association* (AERA). Chicago, IL.

Roth, W.-M. (2001). Gestures: Their role in teaching and learning. *Review of Educational Research*, *71*, 365–392.

Vygotsky, L. S. (1978). *Mind in society*. Cambridge, MA: Harvard University Press.

Zimmerman, H. T., & Land, S. M. (2014). Facilitating place-based learning in outdoor informal environments with mobile computers. *TechTrends*, *58*, 77–83.

Zimmerman, H. T., & McClain, L. R. (2014). Exploring the outdoors together: Assessing family learning in environmental education. *Studies in Educational Evaluation*, *41*, 38–47.

11 The Desired Role of Scientists in Bringing Authentic Scientific Practices Into Classrooms

From Post–Cold War Reforms to Next-Generation Scientists

Asli Sezen-Barrie

One major movement in scientist and school partnerships evolved during the social and political milieu of the post–World War II, Cold War war era in the US. The "scientist in the classrooms" project was a response to an extensive contribution to science during this time. This project failed in its primary goals about raising a population with an increased general understanding of scientific knowledge and inquiry process, but it initiated scientists' contribution to school science and increased the funding to support science education. In this chapter, I explore the long-term impacts of the resulting scientist-centered and top-down curriculum developed during the post–Cold War era. In doing so, I use examples of recent projects to discuss the merits of scientist-school partnerships on creating effective learning environments under four headings: (a) responding to challenges in quality of teachers' scientific knowledge; (b) creating opportunities for meaningful science experiences and fostering authentic science practices; (e) creating equitable teaching environments by community outreach in diverse environments, showing role models, and accessing remote locations; and (f) improving learning sciences research. Finally, lessons learned from these projects will be discussed in order to provide future recommendations for scientist-school partnerships.

WHY DID WE BRING THE SCIENTISTS INTO CLASSROOMS?

It has been 60 years since the Soviet Union launched Sputnik and spurred the US to start an initiative for bringing scientists into schools. This chapter looks back in history to reexamine the lessons learned from this reform and looks at the nature of current collaborations between scientists and schools. Based on what has been learned from the recent efforts, I make suggestions for effective scientists-school collaborations. Even though the Sputnik era reform did not accomplish all of its goals, the influence of the programs developed at the time can be seen today in the norms of content and school science. It is possible to trace the roots of ideas that framed the studies in science education, including the instructional role of laboratories, discipline-centered inquiry, and attention to the nature of science. In his talk at the fortieth anniversary of the launch of Sputnik, Bybee (1997) emphasizes that reforms cannot be evaluated as fail or pass, they move us forward but they have weaknesses. In this chapter, I describe the strengths and weaknesses and then look at the state of current scientists-school partnerships.

A curricular reform had already started in the US before the launch of Sputnik. During that time, the progressive education movement drawing from Dewey's philosophy of education was criticized for not being challenging and distorting facts. Major curricular reforms were seen in every discipline of science. Some of these projects were PCCS (Physical Science Study Committee), Chem Study (Chemical Education Materials Study) BSCS (Biological Sciences Curriculum Study), ESCP (Earth Sciences Curriculum Project), ESS (Elementary Science Study), and SCIS (Science Curriculum Improvement Study). All of these projects shared a common goal to bring scientists into classrooms to have a stronger science education community.

The launch of Sputnik by the Soviet Union on October 4, 1957, accelerated this educational reform and increased federal funding in the US. The National Science Foundation (NSF), Carnegie Corporation of New York, and Rockefeller Brothers Fund lead the financial support for new curriculum efforts. Through these funds

> scientists are drawn in the fray, and that their presence alone helped to legitimize science education reform and elevate its status in the eyes of the public (or at least legislators and the news media). And it turned out that the involvement of those leaders was more then symbolic, for they brought fresh ideas and new leadership energy to the challenge.
>
> (Rutherford, 1997, p. 5)

Moreover, the curricular project gave scientists and teachers a chance to learn about each other's fields. Before then, scientists knew little about pre-college education and teachers did not have an established understanding of scientific content and practice. Despite these drawbacks, the curriculum project was a good start where scientists and educators worked together to develop and field test educative tools.

Despite these successes, the reform didn't achieve all of its educational outcomes, and one important reason was the dominance of scientists over teachers. Rudolph (2002) discusses this irony in his book *Scientists in the Classroom: The Cold Reconstruction of American Science Education*:

> It's ironic that the scientists, while espousing the values of intellectual freedom and creativity, the importance of reason and deliberation in society, worked, in many ways, to routinize and restrict the intellectual role of the teacher through careful construction of their curricular

systems, even to the extent of replacing the teacher completely in some instances as in the film program of PCCS.

(p. 195)

Thus one lesson learned from the Sputnik reforms was that teachers' roles and ideas should be considered seriously in developing new curricula and implementation of new techniques and methodologies in the classrooms. The second important lesson learned from the Sputnik reform was that teachers did not have enough professional development to understand the new content and pedagogy of the revised curriculum. Along with critics from their school system or the culture, teachers resisted change. The third lesson learned was that the reform movement created inequality, as the focus was only on the specific groups of students who were good at science and mathematics and showed promise to continue college education. The curriculum materials developed during that time were not appropriate for average or disadvantaged students.

Exclusion of the science and mathematics education community was yet another weakness of the reform. Teacher educators, science education researchers, and the public were not involved at the initial stages of the movement. The extended communities of teachers were important in the implementation of reformed curricula. For example, teachers needed to attend workshops arranged by science educators who were not main decision makers during the reform. In addition, the time frame of the project was not enough to achieve widespread and sustainable programs leading to meaningful learning in science classrooms. The Sputnik reform was extensive, but this change did not have time to become institutionalized. After the initial push was over, educators tended to go back to traditional practices with which they were more familiar and comfortable.

Although the scientists-school partnerships were implemented following World War II, scientists' involvement in education continues to this day and seems now to be promising to increase the quality of science education in schools. The efforts from Sputnik era have changed over the years and extended their impact to other countries. The neglected role of teachers has been recognized by efforts to support their understanding of conceptual and epistemic aspects of science. The projects have aimed at including all students instead of the elite, and equity became a critical component of scientists-school partnerships. Moreover, science education researchers have taken part in such collaborations and have worked toward groundbreaking research in learning sciences. The next section will use some cases to explain the current state of scientists-school partnerships in order to gauge future directions.

SCIENTISTS-SCHOOL PARTNERSHIPS TO RAISE THE NEXT GENERATION OF SCIENTISTS

Recent research in science education focuses on students' understanding of scientific concepts in harmony with epistemic and social goals. This requires teachers to integrate scientific practices such as "developing and using models" and "engaging in argument from evidence" effectively into the sociocultural environment of their classrooms. Current collaborations between scientists and teachers are expected to help achieve these goals by (a) improving learning sciences research, (b) responding to challenges in the quality of teachers' scientific knowledge, (c) creating opportunities for meaningful science and authentic science experiences, and (d) creating equitable teaching environments.

Improving Learning Sciences Research

In this section, I look at how scientists' involvement can strengthen learning sciences research that will hopefully inform the instruction in K–12 classrooms. Here I focus on two areas of research: making scientific discourse relevant and learning progressions. In doing so, I take a cultural-historical activity theory (CHAT) stance to model how scientists' and teachers' communities of practice intersect at shared goals.

Making scientific discourse relevant. We can explain the importance of using students' daily life experiences by distinguishing spontaneous and scientific concepts. As spontaneous concepts (e.g., floating in the pool) can be grasped in everyday life, students enter into their classrooms with a certain set of scientific vocabulary that they learned from their everyday experiences, observations, or conversations. In relation to those naïve ideas, scientific concepts (e.g., density, buoyancy) are constructed by and can be learned in the social milieu of scientific communities. Therefore, scientific concepts will make more sense to students if they are framed around what they already know.

Here I use an activity, from the Astronomical Society of the Pacific, on scaling the distances between planets to show how scientists can be facilitated to work on students' ideas of scale models and introduce them to space units. This activity is prepared for scientists to use as a classroom activity and to facilitate the elementary schoolteachers who do not have a background in space science. Scientists are encouraged to start talking about scale models through dolls and toy cars, which are familiar objects to students' daily lives. Then the activity uses a one-meter strip of a register tape and marks the location of planets by using stickers to model their distance from the sun. Finally, the scientists relate this one-meter strip to a distance of 40 astronomical units (AU) to help students start using a new concept (Astronomical Society of the Pacific, 2012).

Learning progressions. Learning progressions (LPs) are "descriptions of the successively more sophisticated ways of thinking about an idea that follow one another as students learn" (NRC, 2007, p. 48). The boundaries of this successive progression are called "lower anchor" and "upper anchor." Lower anchor represents the knowledge children bring with them to their school environment and upper anchor represents the knowledge and practices students should be expected to have at the end of their learning progression. The upper anchor can be grounded in expectations for college readiness. Collaboration with scientists in determining the upper anchors of LPs are valuable to be able to determine what society needs to know about a core scientific idea.

An example of partnering with scientists to develop a LP framework is seen in an NSF-funded two statewide project MADE CLEAR (Maryland and Delaware Climate Change Education Assessment and Research). Learning scientists in the projects aim to develop a LP framework for climate change focusing primarily on local issues, such as sea level rise. Although the team uses the recent science education standards documents as a guide to determine the levels of LP, these levels are unpacked with the help of scientists in the project team. Scientists either attend to monthly meetings or arrange a one-on-one meeting with the science educators to provide their insights. Scientists also help to formulate or revise assessment tools to understand if students successfully reached a certain level of the learning progression.

Negotiating between communities of scientists and teachers. In this section, my goal is to focus on the larger communities of scientists and teachers and how this interaction may impact scientists and school partnerships. The explanation here will draw from the sociocultural theories of learning and uses Engeström's (1987) third generation of activity theory. This theory uses activity as the unit of analysis considering a triadic relationship among subjects of the activity, objects that subjects want to reach, and social, cultural, or historical tools (mediating artifacts) that influence the way that subjects reach their objects. Engeström added rules, division of labor, and community. This third generation of CHAT has been used in the fields of teaching and learning.

The third generation of CHAT focuses on the boundaries and transitions at intersections on the objects of the two related activity systems. According to him, activity systems gain more explanatory power when we look at the "transitions and boundaries between the activity system and the actions it generates on the one hand, and between the field of interconnected activity systems in which it is located" (Engeström, 2008, p. 8). Since then, numerous studies used the theory to explain how people navigate between different activity systems or communities. For example, Roth (2007) used the third-generation activity system to understand how attending different activity systems in an environmentalist movement mediate the identities of the middle school students (subjects of the activity).

I use the third generation of activity theory here to understand the multiple dimensions of scientists' and teachers' worlds and to help us understand what teachers and scientists need to know about each other's activity systems. As is seen in Figure 11.1, teachers (left triangle) and scientists (right triangle) share a common objective of "advancing public understanding of science." However, each group can have distinctive means of reaching the same objective because of the difference in their backgrounds, experience in

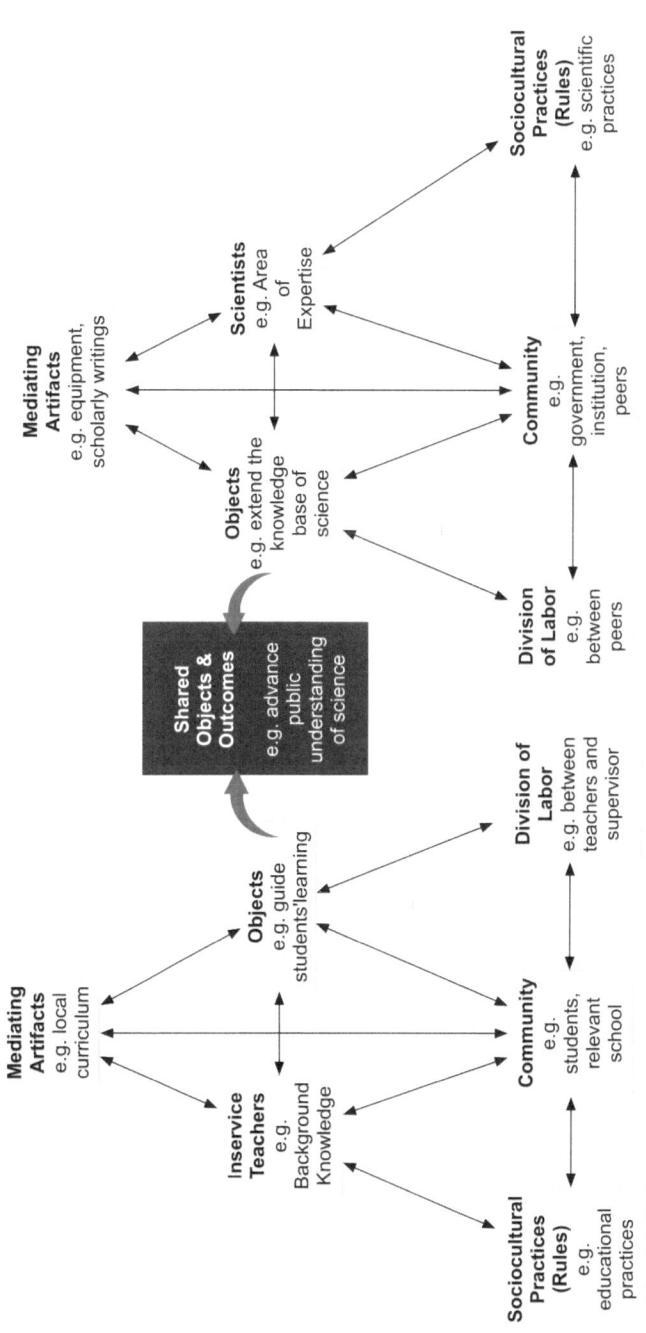

Figure 11.1   Cultural-historical activity theory model of teachers and scientists communities.

using different tools, and how they perform their daily work in their own specific communities governed by peculiar rules.

Thus knowing a scientist or teacher as an isolated individual might not be enough for an effective collaboration. While encouraging such collaborations, funding agencies should create opportunities for teachers and scientists to get familiar with each other's work environment. For instance, short-term sabbaticals can be provided to teachers and scientists to allow them spend time with or work in each other's workspaces.

Framed by the theoretical underpinning of the learning sciences research, the following subsections explain how scientists could effectively contribute to science instruction in K–12 classrooms.

Responding to Challenges in Quality of Teachers' Scientific Knowledge

The recent goals of science education require teachers to demonstrate the ability to not only master the science and engineering content at the grade level they teach but also to facilitate appropriate and effective discourse and argumentation with and among students. However, teachers have a hard time finding high-quality teaching materials, or even to evaluate the validity of resources, due to their lack of background in a certain scientific discipline. A case where scientists help to solve these problems is through a NASA (National Aeronautics and Space Administration) and NSF-funded project ICEE (Inspiring Climate Education Excellence). The ICEE project develops web-based modules to support teachers' understanding of climate change. These modules are prepared with scientists and are tested at workshops with teachers to ensure their quality. The project is currently developing podcasts for teachers on climate science to support teachers' self-directed learning of the content.

Such self-directed learning experience might also help teachers to improve their conceptual understanding of the disciplinary content area. There is extensive evidence from research that teachers' confidence in their conceptual understanding of scientific concepts can help them recognize and use students' ideas, thus making science relevant in the classroom. This understanding is uncommon, however, as teachers traditionally learn scientific content through lectures and have limited exposure to the epistemic practice of science. Learning scientific content through practice helps move away from relying on drilling the facts. To respond to this challenge, every summer the U.S. Space and Rocket Center (USSRC) offers training for teachers at its space camp. During the space camp, math and science concepts are taught using authentic astronaut training simulators. During these simulations, teachers wear real flight suits, use a mock control room to understand the process of space mission control, and go through activities like the "aviation challenge" to learn how concepts of physics can be applied to space science. Such collaborations to increase teachers' scientific knowledge also create venues for their students to learn about current scientific research findings, such as updated ice melting rates in Antarctica.

Creating Opportunities for Meaningful Science and Authentic Science Experiences

During this new reform era, teachers also need to create meaningful and authentic science experiences in their classrooms. To be able to understand all aspects of scientific research and how scientific practices are interpreted differently in various disciplines of science, teachers need firsthand experience. The Cooperative Institute for Research in Environmental Sciences at the University of Colorado Boulder provides opportunities for middle schoolteachers to travel to conduct research in Antarctica with graduate research fellows. One middle school teacher, Ian Schwartz, who used this opportunity to travel during November 2009, said that he could now make science relevant to his students by talking about his experience in Antarctica. Not only did he come back with stories for his class, Ian is also now experienced with different aspects of scientific research. One take-home message for him about conducting science was how little pieces of information found through research can be tied to a larger project to answer a bigger question in science. Another important learning experience for Ian was to recognize the logistical component of scientific research and the fact that what is needed is not always clear or available. He observed how scientists have to solve problems related to equipment (or lack of) during their research. Such problems can occur in science classrooms, and it is important that teachers know how to modify their activities without leading to new misconceptions.

Another example of how scientists can support teachers' understanding of authentic practices of science is when teachers learn to interpret sea level projections data in a summer academy designed by scientists, educators, and researchers as a part of an NSF-funded two statewide project MADE CLEAR (Maryland and Delaware Climate Change Education Assessment and Research). In this activity, teachers learn to use the equipment, how to make notes of their measurements, and how to make projections for the future of the environment. The involvement of the ocean scientists helps teachers to gain perspectives on the domain-specific nature of scientific practices. This includes how to interpret the data for the future of the local environment; that is, how will houses by the ocean be affected?

Creating Equitable Teaching Environments

Scientist and school partnerships have helped create equitable science classrooms by (a) encouraging community service in highly diverse neighborhoods, (b) collaborating with successful scientist role models from underrepresented groups in science, and (c) utilizing technology to connect scientists to remote locations. The National Science Foundation (NSF) in the US continued its efforts to encourage scientists' involvement in educational activities through "broader impact" requirements of grant proposals. A recent goal on improving diversity in STEM education is developing to include all minority groups. During 2012–2013, the funding is provided in programs such as the Participation and Advancement of Women in Academic Science and Engineering Careers (ADVANCE), Historically Black Colleges and Universities Undergraduate Program (HBCU-UP), and Louis Stokes Alliances for Minority Participation (LSAMP).

Yet another challenge that scientist and school collaborations faced was traveling to remote locations. Advancements in technology offered solutions to this problem. A nationwide project, Scientists in Schools (SiS), in Australia showed an example of how scientists collaborated with teachers who are working at remote locations. Since July 2008, project officers were hired to match scientists and teachers and monitor ongoing activities and help arranging meetings. SiS aims to create ongoing and sustainable partnerships between scientists and teachers. An important goal of these partnerships is to "increase scientists' engagement with the broader community, thus raising public awareness of their work and its social and economic importance" (Howitt, Rennie, Heard, & Yuncken, 2009, p. 35). This broader community includes remote locations, which are harder to commute to for scientists. The following case describes how a scientist and a remote school successfully achieved an ongoing partnership.

Diana, a biological scientist, is partnered with Lorraine who is the principal of a small remote school in Queensland, Australia, that teaches kindergarten through seventh grade. Diana's research is on sheep and cattle, and thus she felt her knowledge and research could provide local connections to the students in this school. "She sees her involvement in the project as a means of encouraging students to start asking questions and observing the world around them, as well as seeing how research supports industries in their region" (Howitt et al., 2009, p. 38).

Lorraine and Diana established a method of communication through teleconferencing, email, and regular mail. They established an agenda for the school to work toward improving students' language of science and addressing students' misconceptions on stereotypes of scientists. On their teleconference, the teachers and scientists developed activities that focus on students' observing, describing, and then explaining scientific phenomena. An example activity is when

> the students have observed and drawn spiders in detail. In order to identify the spiders a photo would be sent to Diana. The students were encouraged to ask questions about their work. Any questions the teachers did not feel confident to answer were emailed to Diana. The teachers were also encouraged to email Diana about any science queries they had, as well as just to bounce ideas around. This electronic exchange of

information has helped break down the distance barrier and remove the stigma attached to communicating with a "scientist," and allowed the teachers to feel more comfortable working with Diana.

(Howitt et al., 2009, p. 38)

Collaborating with scientists is helping us fight against inequality problems where not every student has a chance to have an exposure to best practices in science education. Through recent efforts to engage scientists in under-represented and underserved communities, we are one step closer to solving a major problem in education.

RECOMMENDATIONS FOR FUTURE SCIENTISTS-SCHOOL PARTNERSHIPS

The cases described in the previous sections show us only a few of the numerous ways scientist and school collaborations can enhance science education. Drawing from the successes of these cases, we can consider the following recommendations for the future scientist-school partnerships.

One reason for the failure of the Sputnik reform was not including teachers with their sociocultural environment and constraints. Although recent efforts have improved in this respect, there is still a long road that needs to be traveled. Visions, a clear purpose, and goals should be developed collaboratively between scientists and teachers. The activities should be feasible to classroom culture that is constrained with a community and set of sociocultural rules and resources. To achieve this, we should not only anticipate that teachers should think like scientists but scientists should have an improved understanding of complex teaching practices that include identity of the students in the classroom, effective teaching strategies, and the characteristics of curriculum they are bounded with.

To develop a better understanding of teaching and teachers, scientists should be open to understanding their own personal theories and recognize biased assumptions. For example, the codirector of Caltech precollege science initiative, James Bower, a biological scientist, explained the biases he had before starting their partnerships with schools in California as the "myths" of science education reform. These myths are (Bower, 2005, pp. 1–9):

1. The problem with public science education is that a large percentage of teachers are incompetent
2. Teachers are under motivated to teach science because they do not understand how exciting it is
3. The primary reason teachers do not teach science well is a lack of science content knowledge
4. Supplemental teacher training is necessary because too few teachers, especially in the early grades, have been required to take science classes in college

5. The key to scientist involvement with teacher training is to provide complex information in as digestible a form as possible
6. The problem with science education is a lack of good curriculum and therefore we must develop it
7. One reason to develop a new curriculum is to introduce modern scientific techniques derived from current laboratory experiments
8. Training a few highly motivated teachers will produce "trickle down" reform when they return to their schools
9. If teachers are motivated enough during training, they will find a way to obtain the material necessary to teach science in their classrooms
10. Reform can be accomplished with existing resources if they are simply allocated more efficiently

These ideas usually stem from scientists' own experiences in science classrooms and can interfere with developing effective sustainable collaborations.

The evaluation of the nationwide Scientists in Schools (SiS) project in Australia showed that an important reason for not sustaining scientists and teachers collaborations was the difficulty of traveling to the schools, especially to the remote schools. Scientists' change in housing during the time frame of the project created further challenges to sustain meetings. Emails and teleconferences were used in the case described between a biological scientist and a remote school principal in Australia. However, scientists should be cognizant of the resources of the schools they are working with. For an effective collaboration, the school should provide reliable and easy to operate equipment. Technology can be disruptive if not planned carefully.

Another point made in the SiS project was that incompatibility with a partner is one of the main reasons for termination of the project. Therefore SiS hired project officers who worked with teachers and scientists to find a good match by considering the specific needs of schools and scientists. These project managers monitored the progress of the activities and provided managerial support.

CONCLUDING REMARK

As children spend considerable amounts of their time in school, our goal in science education should be to connect the science our students experience during their everyday lives to our teaching. The goal is no longer to just recognize the elite and the most capable to raise a few great scientists. Science education has an important aim to improve *all* students' way of communicating their everyday science to be able to make informed decisions on socio-scientific issues. Scientists can carry an important role in this mission. In this chapter, I aimed to explain how scientists were drawn into classrooms to help the public understand their work and through the years their desired role has extended to areas of equity and learning sciences research.

In doing so, I emphasized the importance of hearing teachers' voices at every step of the collaboration. Finally, special attention was given to the suggestion that sustainable partnerships can benefit from teachers and scientists knowing each other's sociocultural environments.

REFERENCES

Astronomical Society of the Pacific. (2015). *Pocket solar system.* Retrieved from: https://astrosociety.org/wp-content/uploads/2012/09/PocketSolarSystem.pdf (accessed May 13, 2015).

Bower, J. M. (2015). *Scientists and science education reform: myths, methods, and madness. Resources for involving scientists in education* (RISE). Washington D.C.: National Academies of Science. Retrieved on December 15, 2015 from: http://www.nationalacademies.org/rise/backg2a.htm.

Bybee, R. W. (1997). *The Sputnik era: Why is this educational reform different from all other reforms?* Paper presented at the symposium of the Reflecting on Sputnik: Linking the past, present, and future of educational reform, the center for science, mathematics, and engineering education, The National Academies, Washington DC.

Engeström, Y. (1987). *Learning by expanding: An activity-theoretical approach to developmental research.* Helsinki: Orienta–Konsultit.

Engeström, Y. (2008). *From design experiments to formative interventions.* Retrieved from www.fi.uu.nl/en/icls2008/901/paper901.pdf (accessed July 14, 2009).

Howitt, C., Rennie, L., Heard, M., & Yuncken, L. (2009). The Scientists in Schools project. *Teaching Science, 55,* 35–38.

National Research Council (NRC). (2007). Learning progressions. In R. A. Duschl, H. A. Schweingruber, A. W. Shouse (Eds.), *Taking science to school: Learning and teaching science in grades K-8,* (pp. 213–250). Washington, DC: The National Academies Press.

Roth, W. M. (2007). The ethico-moral nature of identity: Prolegomena to the development of third-generation cultural-historical activity theory. *International Journal of Educational Research, 46,* 83–93.

Rudolph, J. L. (2002). *Scientists in the classroom: The cold war reconstruction of American science education.* New York, NY: Palgrave.

Rutherford, F. J. (1997). Sputnik and science education. Presented at the symposium *Reflecting on Sputnik: Linking the Past, Present, and Future of Educational Reform.* Washington D.C.: National Academy of Sciences.

Part III
Universities and Informal Science Intersections

The purpose of this section is to explore the links between universities and informal science programs/approaches, and to examine the role of informal science institutions and contexts on teacher learning and development. Collaborations between universities and informal science already exist in many forms. For example, at the University of Victoria, the POLIS Project on Ecological Governance (http://www.polisproject.org/) works, among others, with watershed boards and environmental activists toward more sustainable water management; and Wolff-Michael Roth worked with local environmentalists in the pursuit of increasing the health of a local watershed and increasing community awareness through children's participation in the activities of one environmental group (e.g., Roth & Barton, 2004). There are many opportunities for learning about science in relation to other fields; and such learning opportunities accrue to teachers and teachers in training as much as to their students. The focus in this section is on university informal science collaborations in the context of teacher preparation, teacher training, and teacher professional development. The six chapters included in this section present an array of conceptual frameworks and provide examples of intersections between university programs and informal science institutions, contexts, and informal science approaches to teacher learning and development. These examples are situated in a variety of geographical and cultural contexts and offer useful insights about the significant role of informal science approaches to teacher preparation, teacher training, and professional development.

The chapters of this section show how informal science approaches can play a vital role in teacher learning and development, and provide evidence that informal science contexts offer a set of unique advantages to teacher preparation and training for various reasons. First, informal science environments offer motivating structures for learning to teach and provide opportunities to practice science teaching in "safe" environments, which, in turn, supports teachers' self-efficacy. Second, informal science environments offer opportunities for learning to teach science through inquiry-based activities in environments that are rich in resources. Third, informal science environments offer unique opportunities for developing content and pedagogical knowledge for science teaching. Last, and perhaps most importantly, informal environments can support teachers in developing understandings about the nature of science, the relationship of science to society, scientific inquiry, and the work of scientists (Avraamidou, 2014). As a matter of fact, research findings provide evidence that engagement in informal science supports the development of teachers' science content knowledge, teachers' pedagogical content knowledge, teachers' development of positive attitudes toward science and science teaching, and teachers' understanding of science teaching and the role of informal science approaches to science teaching and learning. The chapters that follow offer further evidence to the aforementioned and hence provide significant contributions to existing literature documenting the value of greater complementarity between the formal and informal sector and, specifically, the importance of partnerships between university programs and informal science approaches to teacher learning and development.

Karen Knutson and Kevin Crowley, in chapter 12 entitled "Collaborating Across the University/Informal Boundary: Broader Impacts Through Informal Science Education," provide an analysis of the ways in which cultural differences across institutions influence the development and implementation of offerings for learning science. Their investigation focuses on five university-based centers funded by a national granting scheme. The offerings constitute a wide range of possibilities with an equally wide range of target audiences. There are also different ways in which the scientists involved collaborate with other partners, which include (a) handing off the implementation to the educators in the out-of- or after-school settings, (b) training and supporting the scientists involved in designing and delivering outreach activities, and (c) collaborating in professional development. The authors conclude that the nature of the educational experiences varied considerably despite the common orientation of all projects to the same call. Most importantly, perhaps, the potential for sustainability also varied.

In "Museums as Sites for Learning the Art of Education" (chapter 13), David Anderson describes the capacity and power of museums as places where those who aspire to become educators can learn the "art of education" and the skills to become effective educators. The author provides evidence of the impact of practicum-based teaching experiences on educators' philosophy and pedagogy, at the University of British Columbia (UBC) in Vancouver, through two cases: (a) the case of graduate-level students practice within the UBC masters of museum education (MMEd) degree program and (b) the case of K–12 schoolteachers' professional development and training in the context of a practicum teaching program hosted and facilitated in local museums as part of the bachelor of education degree program.

In chapter 14 ("Breaking Dichotomies: Learning to Be a Teacher of Science in Formal and Informal Settings"), Preeti Gupta, Cristina Trowbridge, and Maritza Macdonald describe how a masters of arts in teaching program situated at the American Museum of Natural History in New York City supports aspiring teachers in developing their identities as teachers of science, as people using the affordances of a museum (e.g., objects, exhibits, educators) to ultimately engage school learners. The program, an earth science teacher preparation program, is designed as a 15-month residency program that feeds into a two-year induction program. It consists of two museum residencies and two school residencies. The authors describe the theoretical underpinnings of the design of the program and provide evidence from participant's vignettes to claim that the program creates structured and scaffolded approaches that support participants in shaping their identities as science teachers.

Chapter 15 entitled "City-as-Lab Approach for Urban STEM Learning and Teaching" provides an example of an approach to urban place-based science teaching to support teachers in constructing placed-based identities and enacting place-based practices. In this chapter, Jennifer Adams, Eleanor Miele, and Wayne Powell describe the design of an earth science teacher certification program in Brooklyn, New York. The design of this approach, as the authors describe, centers around three key principles: (a) a thematic approach to integrating community-based resources, (b) authentic science experiences, and (c) collaborative team-teaching and peer-learning. In describing the rationale and theoretical underpinnings of each of these key principles, the authors provide anecdotes and empirical evidence of the impact of City-as-Lab on teachers' development of science teaching identities.

In "The Promise of Collaboration: Classroom Teachers and Use of Informal Science Education Resources," James Kisiel examines, through the lens of identity, the understandings as well as personal experiences of teachers who are able to successfully work at the boundaries of the formal and informal settings. Three informal science institutions (i.e., an aquarium, a science center, and a botanical garden) in Southern California served as focal points for the investigation. Elementary and secondary teachers who had engaged in some way with each institution—e.g., through a field trip, professional development program, and outreach program—participated in an exploratory study investigating the emerging characteristics of these teachers as well as the experiences that were critical on the development of their understandings about informal science.

Framed within the construct of *identity*, Lucy Avraamidou argues in chapter 17 ("Stories of Self and Informal Science: Tracing Preservice Elementary Teachers' Identity Work Across Informal Science Experiences") that as teacher educators aim to prepare high-quality teachers, they ought to examine how beginning elementary teachers construct identities for science teaching outside of school/university and what kind of informal science experiences throughout their lives are critical in shaping their identities. In the first part of the chapter, the author provides a set of conceptual and empirical underpinnings on science teacher identity and informal science to construct an argument about the value of using identity as a lens for examining teacher learning and development. Following that, she presents empirical evidence from a case study, carried out in Cyprus, of the personal stories of two preservice elementary teachers and illustrates how specific, informal science experiences throughout their lives shaped their identities.

In conclusion, therefore, this part III articulates possible links between (a) teacher preparation, teacher training, and teacher professional development and (b) informal science institutions, informal science programs/contexts, and informal science approaches to teacher learning and development. As such, the section provides a basis for conversations aligned with the evolving role of informal science to teacher learning and development, and visions for reform in science teacher preparation, among researchers, teacher educators, informal science educators, and science teachers.

REFERENCES

Avraamidou, L. (2014). Developing a reform-minded science teaching identity: The role of informal science environment. *Journal of Science Teacher Education, 25*, 823–843.

Roth, W.-M., & Barton, A. C. (2004). *Rethinking scientific literacy*. New York: Routledge.

12 Collaborating Across the University/Informal Boundary

Broader Impacts Through Informal Science Education

Karen Knutson & Kevin Crowley

In this chapter, we present the ways in which institutional cultural differences impact the development and implementation of learning activities in informal settings. Five university-based centers for the study of chemistry worked with informal learning professionals to re-envision educational and public outreach activities about science. The projects were part of a broader effort to catalyze new thinking and innovation in informal education and chemistry centers. The set of projects illustrates the broad possibilities for informal learning settings, with projects targeting diverse audiences with a range of experiences, including an interactive exhibit at a major science center, activities for an small drop-in science club for disadvantaged teens, curricula for organized after-school and summer camp programs, audio science stories for a general audience, and a fellowship training program for informal educators and scientists. We highlight the ways in which professionals working in universities and informal settings structured their collaborations and reflect upon the conditions that led to success on a range of dimensions.

Historically, outreach has been a part of large multi-investigator science centers funded by the National Science Foundation (NSF) in the US. In addition to their core scientific research missions, these centers conduct education and public outreach activities that take many forms, including both short and longer-term programs or one-day events. Centers might fund internship programs for undergraduate or graduate researchers, providing a structured training opportunity for young scientists. High school students might have a weekend-long program to learn about science in a specific area. These kinds of outreach activities are often organized around demonstrable outcomes that positively impact the career pipeline for future scientists. These kinds of outreach programs have a formal structure and serve the goals of the higher education system. Centers also develop experiences for general public audiences. These activities might include lectures, participating at tabled events, or, more recently, working at dialogue-based events such as science cafés. In many cases, outreach activities fall to academic staff who do not necessarily have expertise in facilitating these kinds of less-structured and informal experiences or working with public audiences of varied ages and backgrounds (Andrews, Hanley, Hovermill, Weaver, & Melton, 2005).

Whereas public outreach is an activity that may be seen as necessary and desirable, its impacts are harder to document (Brown, Yeung, & Sawyer, 2014; Neresini & Bucchi, 2011). The structure of academic science programs provides no reward system for these endeavors, and scientists often feel that they do not have the time or the skills to implement innovative programs and activities for public audiences.

In 2009, in light of these and other concerns, NSF reexamined its grant criteria and decided that grantees should place more emphasis on the quality and quantity of outreach communication, requiring all research proposals (not just the large centers) to explicitly address the "broader impacts" of research on science, education, and society. Broader impacts criteria include wide dissemination, increasing infrastructure for research and education, reaching underrepresented groups, engaging the public and K–12 audiences, and providing professional development for teachers or early career researchers. Importantly, broader impacts criteria are to be as rigorously considered as intellectual merit during proposal reviews. With its desire to better connect society and science, the new broader impacts criterion is noble, but in practice improving broader impacts and outreach remains difficult (Alpert, 2009).

As an experiment in strengthening broader impacts work, in 2012, NSF initiated a special project for its Centers for Chemical Innovation (CCI) that was designed to encourage mutually beneficial and sustainable collaborations between science, technology, engineering, and mathematics education (STEM) researchers and informal science education professionals. The CCI's are large multi-institutional research centers where scientists work on long-term challenges within targeted areas of chemistry, such as solar energy, chemical evolution or catalysis. CCI's are heavily university based, but they may partner with researchers from industry, government laboratories, or international organizations. CCIs are "tasked to integrate research, innovation, education and informal science communication," and it is hoped that by bringing together researchers with shared and complementary interests, a culture of risk-taking and innovation will result (NSF, 2012).

The Informal Science Education initiative was created for CCIs to encourage collaboration between the centers and informal science professionals in order to develop awareness of and expertise in working with informal science education (ISE) practices and to ultimately improve the quality and expand the scope of education and outreach activities provided by these centers, and scientists more generally. The funding for these collaborations was framed as seed money, with the goal of catalyzing new collaborations with informal educators to experiment with novel and innovative approaches. As they searched for informal education collaborators and formulated project ideas in response to the grant, many of the centers made use of the resources of the Center for Advancement of Informal Science Education (CAISE), including consulting with CAISE staff and using the materials on its website (http://informalscience.org). The centers had five different visions of project type, audience, and manner of working with partners. The experiences produced included an interactive museum exhibit, after-school science programs, audio spots, activities for a drop-in science club, and a professional development environment.

THE RESEARCH STUDY

Our study of this funding experiment explored project collaboration and the issues and challenges faced by new partnerships, as well as the impacts and innovative aspects of the projects. Data collection involved structured interviews with staff in each of the projects, including center scientists, center education and public outreach coordinators, and informal science professionals. Interviews were transcribed and coded for themes and patterns. We also analyzed documents and products from the partnerships. Follow-up interviews with each project team were conducted six months later to document project progress and clarify our emerging themes.

Our analysis focused on how university-based scientists approached the challenge of collaborating with informal learning partners, what they learned from the collaboration, and various ways we might consider success in light of varying measures of impact, innovation, and sustainability. Specifically, our guiding research questions were:

1. What was the nature of the relationship between partners before the partnership and what is it like afterward?
2. What impact did the partnership have on each partner (e.g., changes in understanding of each other's culture, success of the product/services developed during the supplemental grant)?
3. What is the probability that the relationship will be sustained and how will it be the same—or different as it moves forward?
4. What are the characteristics of the partnership(s) where the greatest understanding of each other's culture occurred? What are the characteristics of the partnership(s) where the probability of sustainability is the greatest?

Overview of the Five Cases

Exhibit. One center developed an interactive exhibit that would fit within an existing exhibition on energy at a large science center in an east coast U.S. city. An interactive computer-based touch table was designed to communicate the role of catalysts in speeding up reactions essential for creating products from petrochemicals. The activity allows users to create chemical pathways that take fossil fuels and turn them into molecules responsible for common items that use petrochemicals, such as aspirin and lipstick. The project was envisioned as a way to reach the general public and develop an awareness of the vast and significant ways that we use fossil fuels. In terms of process, exhibit designers at the museum were connected with a CCI member, an assistant chemistry professor, and students at a nearby university. The academics were invited to brainstorm meetings with museum staff. Students then created prototypes for the exhibit as part of their classwork. Students' ideas were important fodder for the exhibit design team, and their professor continued to provide input into later stages of development. Whereas scientists learned something about informal education and museums by working with museum staff, this was primarily a consultative process. The resulting exhibit is engaging and educational, reaches a large audience of adults and children, and relays important science concepts germane to the work of the research center. Whereas the science content communicated is relevant to science of the center, the need to be accessible and interactive somewhat impacted the depth of concepts that could be tackled in this format.

Media. A center in a large southeastern U.S. city partnered with an independent radio producer to develop audio stories aimed at enriching public science literacy. Audio segments used a variety of techniques, including standard public radio narrative style, short scientist-narrated nuggets, and imaginative explorations of key chemistry concepts. Scientists in the center also worked with the producer to develop their communication skills as authors and creators of audio pieces. Live performances that used content from the pieces were also staged. Before this collaboration, the center had a long history of working within the university community on innovative arts-based outreach projects, so the team was excited to try out a new media format. The producer worked with staff to develop stories that could serve as a bridge between public interest and the complex kinds of chemistry concepts that the center focused on. He also offered a workshop for 20 chemistry students on radio production skills in addition to the narrative story crafting skills required for the project. Three professional development events were also conducted at two universities during the course of the project.

After-school. A center in in a west coast U.S. city worked with an on-campus educational resource center to deliver hands-on science programming to engage participants (age 8–12 years old) in STEM activities in after-school settings, as well as during a four-day summer camp for 8–9-year olds. This particular university had an existing educational outreach group that, among other things, was already developing and delivering STEM content and working with teachers. As the ISE partner, staff members from the educational outreach group were central to the project, developing and coordinating the activities with student teachers, and implementing many of the lessons at the club. The outreach coordinator from the chemistry center worked closely with outreach staff and attended nearly every program to see lessons in action. She was the connection to center scientists. One ongoing challenge for the project was to connect the challenging science content of the center to the outreach activities for children of this age.

The team worked hard to find ways to connect the science of the center to hands-on experiences. The initial intent was for the after-school club to take over the implementation of the designed activities, with the team providing the activities and training for instruction for the boys and girls club staff. However, low science knowledge/ interest as well as turnover of staff at the club required that the partnership continue to provide instructors for the activities.

Science club. Another center in a west coast U.S. city created a set of activities for an after-school neighborhood club for local youth. This partnership was very collaborative, with the informal partner working with faculty and postdocs to develop and deliver outreach activities at the club. Workshops were held to help scientists become oriented to the ISE field. Several postdoctoral fellows were closely involved in the process, attending programs at the club, and becoming part of the club's culture. This club is small and intensive, with about 8–10 regular kids, and this provided an opportunity for the scientists to see that they were making some very real impacts on children's lives with science. The team developed 15 modules focused on chemistry. The development process involved creating mind maps for the youth and relating them with topics of scientific research at the center. The first year also involved field trips. For the second iteration, the team decided to focus on a few of the modules in more depth, expanding them to two-month activities each with four sessions. Although general chemistry was covered, there was also a focus on topics that were more closely related to the cutting-edge chemistry research of the center. For a field trip, the club went to a hydrogen filling station. Youth looked at fuel cells and did projects on polymers and visited an advanced lab. Partners felt that the partnerships developed in this project would extend beyond the life of the grant, and, additionally, postdoctoral fellows working on the project told us they are now rethinking their career direction to include a stronger focus on science communication.

Professional development. Finally, a center based in a northwest U.S. city worked to develop a strong partnership between scientists and museum educators to lead to a deeper understanding of each other's practice. Calling themselves a "Partnership for Public Engagement," the project mounted a series of professional development seminars that brought together scientists and informal educators. During the seminars, scientists introduced museum staff to their research interests and research. Museum staff worked with researchers on science communication skills and issues. The cornerstone of the project involved the creation of a fellowship program for student scientists to work on ISE activities with museum staff. This program recruited participants by application and three grad students were selected. During their meetings, museum staff and fellows worked on designing and testing outreach activities based on the scientific research of the center. A template for program design was created and, through joint discussion, museum staff realized that their template required more specificity for use by those outside of the ISE field. These revisions helped the team to better address language issues involved in communicating with a broad range of public audiences. Center fellows engaged in prototyping program designs and found it an enlightening process, helping them to understand firsthand the communication challenges of working with different audiences. Fellows also learned about the particular challenges in developing hands-on activities for informal settings. Museum staff report that they are much more comfortable with the science and the scientists, and fellows love working with the museum. The project planned to continue the fellowship program with other funding and looked for future opportunities to collaborate in other areas.

Structuring the Collaboration

In each case, the center was required to form a new collaboration with an informal learning entity, although the cases varied in the degree to which the collaboration was novel and/or ambitious. We consider the cases first in terms of the three distinct models they took to collaboration: hand-off, training, and collaborative.

Scientists hand off the implementation to informal educators. The hand-off model was most clearly implemented by the museum exhibit and after-school cases. The design and delivery of learning experiences was deemed to be largely the role of the informal learning staff, with center scientists playing the traditional role of outside science content consultants. The after-school case saw very little involvement of scientists in program development. In large part, this was because the center decided to utilize an established university-based outreach group staffed by experienced science teachers; the latter were already developing and running educational programs. There were conversations around how to include cutting-edge chemistry in programs, but, in the end, the experienced science teachers drew upon the kinds of content and approaches that they knew would work for the classroom-like settings in which they would be working. The center staff member tasked with running education and public outreach remained actively involved throughout the project, but the scientists were mostly disconnected from the work.

In the exhibit case, the scientist as consultant was more involved and the relation was prolonged, with scientists participating in multiple design meetings and, through extended exposure to the museum and its staff, learning about informal learning experiences in museums. Once the concept was developed, however, the museum took over design and production details for the interactive, using their usual, well-developed processes. The scientists got a glimpse into the issues of designing informal experiences and grappled with the appropriate level and nature of cutting-edge knowledge to communicate, but, in the end, it was the product that was the main focus. And once the product was completed and on the floor of the museum, the collaboration had run its course. From the perspective of the museum staff, the project represented minimal opportunities for learning and change. The effective use of content consultants and stakeholder co-design are well-established parts of an exhibit developer's typical repertoire of practice.

Informal educators train, scaffold, and support scientists who are involved in designing and delivering outreach mode. In the science club and media cases, the collaboration was structured to change the roles and relationships between educator, scientist, and audience. In the case of the club, especially, there was evidence of high levels of collaboration and learning across the university/informal boundary. Compared with the classroom-like settings of the after-school case, the science club was small, neighborhood-based, and "owned" by local youth who dropped in for science activities. Building from an established trust with the youth, the club director systematically brokered the relationship between the university and the youth, having the scientists become a recurring presence in the club, getting to know the youth, iteratively co-designing activities, and traveling with the youth when they went on field trips away from the club. This was a critical link between scientists and youth—and it is important to note that the link was built on the turf of the youth as opposed to the scientists' turf. Getting out of the university and into the community is essential for the kind of deep and extended impact to which this project aspired. The scientists took on the role of instructors and mentors to an organized group of underrepresented urban youth who had, by being part of the club, already identified themselves as looking for deeper engagement with science. It would have been very difficult for university-based scientists to effectively reach such an audience without leveraging existing structures such as the club.

The media case also provided opportunity for learning and boundary crossing but focused more directly on skill building in the scientists. The collaboration was structured around broad science communication goals, with the producer providing training for scientists in communicating with public audiences through the collaborative development of the audio spots. Reflecting their prior experience working with public audiences and collaborating with artists and producers, the center leadership approached the collaboration with an open-ended and experimental stance. Both the producer and the scientists saw this opportunity as being about more than simply producing the specific content and looked for opportunities to leverage the relationship for mutual benefit. In terms of the content, the media case provided some of the strongest examples of effective informal learning experiences focused on a center's complex and cutting-edge science.

Collaborative professional development model. Aspects of the media and youth club cases reflect a move from training toward collaboration. The last case, the professional development case, is a strong example of a deliberate effort to build a new collaborative relationship. Processes of working together with an audience in view were always in first position for this project, and both scientists and educators judged success primarily on what they were learning as they worked together. Museum staff learned about the science, and scientists learned about educational techniques and practices, and they jointly utilized these skills in prototyping educational experiences for museum audiences. In this case, both the museum and the center were experienced collaborators and had staff on hand who could guide the project through some of the common pitfalls of working across the university/informal boundary. The desire to parlay the seed funding into a larger relationship echoed larger organizational strategies for both the museum and center—we were told in interviews of a regionally shared value for collaboration and how, at the local and state levels, collaborative connections between universities, nonprofits, and for-profits were being actively encouraged and supported. In terms of sustainability, the museum fellows program emerged as an important outcome for the project and provided a mechanism to incorporate informal education experience into the graduate and postdoc training programs.

Considering Success

So in which of these cases did the funder's investment in collaboration and innovation pay off? The diversity of approaches taken in the project makes it difficult to talk about success and impact in a single way; in many ways, success depends on our own vantage point. For example, from the perspective of the larger informal learning field and the funder, we might tend to prioritize dimensions that highlight innovative practice that could be shared and scaled with other informal educators. The funder might also place great value on the number and diversity of audiences served with these initiatives. From the perspective of the centers themselves, we might place value on the utility and sustainability of these specific partnerships. From the perspective of larger communities of scientists interested in broader impacts, success might be defined as lessons learned about the value of having scientists interact directly with public audiences, the institutional barriers and supports for scientists becoming involved in broader impacts, or the role of communicating cutting-edge as opposed to general science in informal settings. Table 12.1 highlights a number of dimensions upon which these project activities could be measured. The columns can be seen as a continuum, but we caution that the continuum does not prescribe one particular ideal outcome. Rather we see this as a matrix that one could use, selecting a range point for each criteria that would result in a program that best suits the desired audience, science, and budget.

Boundary crossing. One metric of a successful collaboration might be to ask how many people moved across the university/informal boundary, and what parts of the project did those people represent? The logic here would be that supporting more boundary crossers would help to create a more broadly shared knowledge in an organization, make the partnership more resilient to personnel turnover, and increase the potential for innovative and sustainable collaboration (Russell, Knutson, & Crowley, 2013). By this

Table 12.1   Dimensions upon which to measure success.

Typical	Innovative
Low number served	High number served
Children	Adults
Short exposure	In-depth exposure
General science	Specific center science
Formal	Informal
Low investment	High investment
Arm's length collaboration	Close collaboration
One time effort	Sustainable
No relationships	Ongoing relationships
Works with existing structures	Initiates change in center and/or university

metric, the media and professional development cases are particular stand-outs, with the ongoing collaborative activities bringing together relatively larger numbers of individuals from across different parts of the centers. In contrast, the crisp hand-off model of the after-school partnership effectively isolated any potential organizational learning and change in the experience of a very few people.

Professional learning. Beyond the number of boundary crossers, one might also ask what people did when they crossed a boundary and then what they took with them when they moved back across to their home context. This metric would favor particularly the science club case, as the scientists who developed modules for the youth were working for prolonged periods in new roles that were far beyond the typical comfort zone of a university-based researcher. The exhibit, media, and professional development examples also would be successful by this metric, as they all involved moments when scientists crossed over to do the work of informal educators to varying degrees.

Numbers served and efficiency. The projects illustrate trade-offs in how they positioned impact to themselves versus their public audiences. The two hand-off projects opted for well-understood genres of work in well-established settings that already had built-in audiences. The projects were inherently low risk and best described as one way, transactional, and business as usual, with resources and specific chemistry content flowing from the center to be packaged by the informal educators, to be delivered to a large-as-possible audience. From that standpoint, both of these projects were successful. The after-school curricula reached many groups of students and the museum exhibit continues to reach thousands of visitors. The products of each of these projects could be fairly easily "shrink-wrapped" and sent to other similar settings for immediate use, increasing their potential impact even further.

Institutional change. Center staff reported a positive impact of the informal education experience on their perceptions of outreach and their desire to be involved in this kind of work in the future. However, we saw more limited change in the ability of centers to support outreach in a new way. In many interviews, staff lamented the fact that they had more ideas for what they might do if they had had more money or time, but the end of the grant would effectively end their ability to work together. Given that these projects were relatively short-term and created for the grant, there were always potential threats for sustainability. Without a strong interpersonal connection between sites, or joint goal setting emerging from a project, it would be hard to sustain collaboration. Additionally, in cases where the collaboration was isolated to one or two key contact persons, there is always the threat of staff turnover. Indeed, three of the centers had significant members of the team leave the group before or near the end of the project. Collaborations that involved creating new institutional processes, shared goals, and involved larger numbers of staff have shown the strongest sustainable elements.

Innovation in informal learning and science communication. Dealing with content that is difficult to communicate was a common concern for all of the projects. The after-school project team was initially interested in activities related to the content but ultimately decided that the needs of the more formal structures of the after-school classroom could not be met without shifting to simpler content. The media project was better able to communicate cutting-edge content, as their format is adult-oriented; but even so, they found it hard to make visible some of the processes that their scientists worked on. Some of the programs that were targeted for school-aged children struggled to make new content that related to the center research mission, whereas others felt that their time was better spent addressing more general science topics, such as lab safety or science inquiry skills—the "basics" that students might be thought to need for success in school-based science instruction.

CONCLUSION

Comparing these five cases of university/informal collaboration is revealing in terms of identifying features associated with innovation and potential sustainability. In some ways, all five collaborations might have been expected to be similar: after all, they were all initiated by large university-based research centers, they all shared similar levels of funding, they all had dedicated staff who were already working on educational outreach, and they all shared the same goals and reporting requirements of NSF funding. Despite these structural similarities, the form of the collaborations, the nature of the educational experiences developed, and the potential for sustainability turned out to be, as we detailed earlier, quite varied.

We conclude by reflecting upon the future of university/informal collaborations, noting current barriers that will have to be overcome in order to achieve successful collaborative ventures. Institutional conditions often work directly against scientists getting involved in this kind of work. We heard in interviews of how the work is often not valued by the university system and made possible only because of the funding. This is despite the fact that there are many scientists eager to participate: The projects we studied were generally able to identify scientists who were very keen and inspired by the work, but even they often acknowledged that this activity may be detracting from their core academic research. This was a particular concern of the graduate students and postdocs, who often told us that they hope to bring their new informal expertise with them and connect it to their jobs as university-based scientists. We will need to do more than expand a scientist's own knowledge and skill of informal learning to see that hope realized. We will need to transform universities and scientific research centers to support in ways that encourage continued exploration of science education and outreach by crossing the university/informal boundary.

REFERENCES

Alpert, C. (2009). Broadening and deepening the impact: A theoretical framework for partnerships between science museums and stem research centres. *Social Epistemology, 23*, 267–281.

Andrews, E., Hanley, D., Hovermill, J., Weaver, A., & Melton, G. (2005). Scientists and public outreach: Participation, motivations, and impediments, *Journal of Geoscience Education, 53,* 281–293.

Brown, E., Yeung, L., & Sawyer, K. (Eds.). (2014). *Sustainable infrastructures for life science communication.* Washington, DC: National Academy of Sciences.

National Science Foundation (NSF). (2012). CCI ISE supplemental funding opportunity. Retrieved on December 15, 2015 from: http://www.nsf.gov/funding/pgm_summ.jsp?pims_id=13635

Neresini, F., & Bucchi, M. (2011). Which indicators for the new public engagement activities? An exploratory study of European research institutions. *Public Understanding of Science, 20*, 64–79.

Russell, J., Knutson, K., & Crowley, K. (2013). Informal learning organizations as part of an educational ecology: Lessons from collaboration across the formal/informal divide. *Journal of Educational Change, 14*, 259–281.

13 Museums as Sites for Learning the Art of Education

David Anderson

Museums, as an integral part of the community in societies across the globe, serve as the repositories and trustees of cultural and historical wealth; institutions of research; destinations of leisure; and sites of education experience for tourists, local members of the community, families, and school groups. Indeed, museums are institutions of both education and learning in service of society; that is, they are sites where educational experiences are designed, mediated, and facilitated, and also places where visitors learn. The focus of this chapter is on the latter role. Museums rarely conceive of themselves as places that afford opportunities to learn how to become an effective educator. Rather, most people in the museum field conceive of their primary role as being the repositories of cultural wealth and perhaps secondary places of scholarly research. However, through the exploration of the cases of museum practicum in two academic programs at the University of British Columbia (UBC), Canada, and their collaborative partnerships with museums in the city of Vancouver, this chapter reveals the considerable power of museums as environments for producing effective and highly skilled educators to the benefit of society.

THE EDUCATIONAL ROLE OF THE MUSEUM

"Museums are primarily educational institutions; what makes them public institutions for the preservation of culture is their educational work" (Hein, 2005, p. 357). Indeed, museums afford opportunities to make connections and meanings between their cultural representations (artifacts, exhibits, and experiences) and their visitors' own cultural worlds in ways that enhance their knowledge via *mediation through artifacts* and *human mediation.*

As the boundaries that once separated museums from other recreational and educational organizations blur, and their role and function as places of education in society increase, so too does the need for skilled professionals who understand the nature of education in these kinds of settings. Hence the need to give thought to "who" within the museum mediates educational experiences and how the museum and its education practices influence and shape the educational philosophy and pedagogy of the mediators themselves.

The term "museum educators" is gaining increased usage in North America and elsewhere within the museum profession. The term is also often employed to describe museum professionals who take on diverse sets of responsibilities and who are trained or skilled to understand the nature of educational practice. The term is used to describe, but is not limited to, those who develop, coordinate, implement, and manage programs for school groups, families, teachers, and the general public. In addition, "museum educators" might include professional designations that involve the creation, development, and nurturing of relationships with community groups to foster accessibility and usage of museums, and to enhance the relevancy and inclusiveness of the people they serve. Further, museum educators might contribute to the conceptualization, design, and development of programs and exhibitions.

Skilled museum educators, who work on the staffs of thousands of museums around the world, play a central role and support formal K–12 education systems and their benefactors. I also argue that, in the case of K–12 school group visitors, the educational experience should also be rightly mediated not only by the museum's educator (either directly or indirectly) but also by the schoolteacher (Anderson, Kisiel, & Storksdieck, 2006). In this sense, both the museum educator and the teacher, ideally, are the co-facilitators of educational experiences for visiting students. However, K–12 teachers are professional educators whose skill sets and training may be argued to be narrowly bound within the confines of a particular mandated curriculum and formal education. Teachers, around the world, are not often educated within their university teacher-trainee programs to know how to practice the *art of education* effectively outside the bounds of the classroom environment. Becoming an effective teacher practitioner requires experience as an educator in contexts beyond the limited confines of school-based classrooms. To become fluent in the *art of education* requires a much broader skill set than classroom-based mediation of learning experiences.

It is with this premise and assumption that I explore the considerable power of museums as environments for producing effective and highly skilled educators by investigating the impact of practicum-based teaching experience on educators. In particular, the nature of such impact in terms of the educators' philosophy and pedagogy that was being shaped by the museums' educational practices in which they were enculturated. The professional development and credentialing of two kinds of educators were explored within the bounds of two different academic programs at UBC and their collaborative partnerships with museums.

CASE #1: DEVELOPING MUSEUM EDUCATORS IN THE MMEd DEGREE PROGRAM

In North America in recent years, there has been the emergence and rise of a small number of graduate degree programs focusing on museum education as a discipline that are philosophically oriented toward the professional accreditation of museum professionals in the field of education. UBC's MMEd is a one-year intensive full-time, or two-year part-time, graduate degree program focusing on the study of education and learning that occurs in a broad diversity of museums and other informal learning contexts. It is different from *museum studies* programs in that it focuses purely on the scholarly study of education as a discipline and not on curatorial practices, such as conservation and preservation. The program aims to develop museum educators to become catalysts for different ways of thinking about the educational roles and potential of museums for teaching and learning. Moreover, one of the major emphases of the program is to wrestle with the significant and complex issues about how museum institutions communicate, interpret, and ultimately educate visitors in effective ways and challenge the adage that one kind of communicative or interpretive approach suits all. At the heart of the issue is the "pedagogy of the museum."

Nature of the MMEd Program and Museum Field Experience

A component of the studies with the MMEd is "Museum Field Experience" in which graduate students participate in about 120 hours of practical experience in museum settings of their own interest and discipline specialization. Because the MMEd is a graduate program open to all subject specializations, it attracts people from the sciences, arts, social studies, and humanities disciplines. As such, graduate students opt to undertake their field experience in eight different museums in the city of Vancouver who partnered with the MMEd program. During the course of the 120-hour field placement, each graduate student worked within their assigned museums' department of education, educational outreach, or visitor service department depending on the museum type. Here they were involved in a wide diversity of tasks including, in-gallery teaching; leading tour groups; developing new educational programs for diverse audiences; conceptualization, design, and development of exhibitions; community liaison; evaluation of exhibition and programs; and so forth. They did these tasks in conjunction with the museum staff and museum teams as part of the normal day-to-day function and operation of the museum itself. Further, the field experience component of the MMEd program was situated about two-thirds into the program, and as such they were readily able to apply much of the museum education theory they were exposed to as part of their university studies to the museum environment in practical ways.

The Impact of Museum Field Experience for MMEd Students

The research approach employed to investigate the impact of the museum field experience was classified as a qualitative, interpretive case study and involved 17 graduate students who had completed their museum field experience in the 2013 academic year. At the conclusion of their field experience, students were interviewed individually to ascertain the impact of their experience on their views about museum pedagogy, visitor learning, and their identities as educators. The interpretations of the data resulted in six themes with supporting excerpts from student interviews.

Rich understandings of how visitors learn in museums. Graduate students developed a rich and detailed understanding of how visitors learn in museum environments, and in particular, the notions that visitors bring with them experiences and prior knowledge, which influences how they behave and how they learn in the museum galleries. As one of the graduate students stated:

> Visitors come to museums with different prior knowledge, which influences their interests and what they attend to in the museum. Also, they continue their learning experience after exiting the museum. Consequently, learning in museums should not be a "product" that we just want to give to the visitors; it should be a process in which the museums play different roles to different visitors at different stages of learning, and with each return visit.

Further, students came to appreciate that "museums play different role to different visitors at different stages of learning." This idea points to profound appreciation that museums are not merely one-off visits during which one acquires all there is to know from the museum. Rather, they ought to be considered multi-visit destination from which multiple learning outcomes can be realized with each visit and return visit.

Appreciation of visitor identity to effectively communicate museum messages. An appreciation of the different learning needs of various kinds of museum visitors, as audience demographic, was a commonly expressed impact of the field experience. Indeed, appreciating the notion that one style of mediation, from programs or exhibits, may not necessarily suit the learning needs of all types of visitor demographics was a common revelation to the graduate students. All students came away from their field experiences with new and profound transformations in their appreciations of different kinds of museum demographics and how to mediate messages for various visitor identities.

> Overall, my most significant learning has been having the experience of working through challenges associated with teaching and learning on the ground—I now have a deep appreciation of how to translate diverse visitor's learning outcomes—from all kind of audiences, the young and the old, families and teenagers—into real-life museum experience exhibits and programs for wide and diverse audiences.

Understanding of the learning dynamic between museum medium and visitor. The role of the museum in communicating its content and messages is not simply a matter of the transmission or portrayal of information about science, history, art, or culture. It is much more complex and requires an epistemology of learning that focuses not just on the object or artifact and its meaning, but rather on both the museum as the mediator of content and on the visitor as learner.

> In the beginning, my practice as an educator in the museum used the transmission-oriented approach—"just tell them the information"—I used to think that this was the best teaching method. It was what we were exposed to in schools, focusing on content to be taught rather than the visitors as learners. The field experience has kept me challenging myself to understand the needs of the visitors and not just on the content and curriculum I want to push. Both are required—understanding of the content of the museum, but also knowledge of the visitors themselves.

The power of social experience on learning in museums. All students developed deep understandings and appreciations of the importance of personal and social engagement with visitors in the museum on promoting rich learning outcomes. This speaks to the importance of being exposed directly to those whom the museum serves.

> I came to appreciate that educational practice in museums allows for an innate flexibility and remains open to the interactions between strangers (visitors) in a setting where the "life" of the exhibition can depend on these communications. I learned how to interact with visitors, how to ask questions that can promote their thinking, and the importance of engaging with my visitors and helping visitors to engage with one another. My field experience helped me to validate the inherently personal meanings visitors collectively construct from any museum experience whether it was intended to be educational or not.

Moreover, understandings of the importance of the social dynamic that exists between visitors sharing the museum experiences on learning in the museum was prevalent among the students' reported outcomes from the field experience.

The power of the emancipated environment on learning in the museum. Following closely from the previous theme was the value and appreciation a number of students developed about the notion of the museum as an "emancipated" learning environment in which visitors were free to question, talk, and exchange ideas with others. The following excerpt expresses this ideal goal of museum education and that museums should strive to create and give permission to visitors to express ideas freely in galleries. Such an emancipatory learning environment has the capacity to enrich the learning experience of museum visitors:

> Education in museums is not only about information, but also about creating a free learning environment. To create a free environment for visitors to exchange ideas; let them know that it is fine to have their interesting conversation with one another in the gallery in their normal voice—this had not been my previous conception before; now it is!

Museum educator identity–educator and learner. All graduate students had transformations in their identities as museum educators as a result of the field experience. Most students went into the experience feeling that their role was "teacher," because they held views of themselves as "museum educators." However, the field experience was for many a somewhat humbling experience that presented numerous challenges for which they had to become "learners" of on-the-job skill and, on many occasions confront the limitation of their own skill and knowledge of museum education practices.

> I transferred my identity at my museum workplace from "a teacher" to "a learner and an educator." The experience of being a learner helps me to have a deep reflection on how to teach others to teach in museums. I felt the pressure of self-imposed expectation that I must know everything. But, in reality—you can't know everything. Even sometimes visitors ask you questions which you don't know the answers to, or you have to develop certain kinds of exhibits or program and you feel you don't know what you are doing.

These key transformations in the student educators' understandings of (a) museum pedagogy, (b) visitor learning via the dynamic between mediation through artifacts (exhibits) and between one another (human mediation) as well as (c) their own changing self-identities as learners and educators where significant epiphanies emerged as a result of the museum field experience and enculturation.

CASE #2: MUSEUMS AS PLACES TO SUPPORT THE DEVELOPMENT OF K–12 TEACHERS

In recent years, there has developed an increasing appreciation of the benefits of preservice teacher education programs collaborating with museums as a means to enhance the pedagogical skill sets of those training to become teachers (e.g., Anderson, Lawson, & Mayer-Smith, 2006). In keeping with the sentiments of this appeal, UBC's teacher education program partnered with the local museums in the city of Vancouver to develop an innovative practicum experience for teachers in training that sought to expand and re/form the traditional definitions of teacher education. The hope was to produce teacher graduates more fluent in the *art of education*.

The Bachelor of Education Program at UBC and the Museum Practicum

Among the teacher education programs at UBC, there is currently a 12-month degree program that enables candidates holding bachelor's degree qualifications in a discipline area (science, arts, social studies) to complete a bachelor of education degree in secondary-school education. The practicum model within this degree program comprises a 13-weeks extended placement designed to help students connect pedagogical theory with practice; provide teaching experiences that provide preparation for a career in teaching; encourage preservice teachers to reflect systematically and analytically upon teaching in a professional and educational community; and create opportunities to plan, implement, and evaluate instruction.

The partnership with local museums was initiated in an attempt to re/form this traditional classroom-only model of extended practicum to one that consisted of a ten-week classroom placement followed by a three-week teaching experience in the museum setting. The rationales for this change were numerous. On the university side, there was a concern that practicum experiences offered in school-based settings were not in keeping with broader definitions of teacher education. Teacher education is about equipping educators with a wide range of skills that can be readily transferable across contexts, inside and outside of school settings. Such education and training should provide preservice teachers exposure to and opportunities to practice in a wide array of learning environments.

Preservice teachers were assigned to partner museums as a function of their professional interests. During their three-week museum practicum, preservice teachers were involved in both teaching and developing materials for the curriculum-based school programs offered by each of the museums for K–12 students. Teaching in the museum comprised multifarious activities, including: (a) meeting, greeting, and organizing school groups; (b) introducing and concluding the program activities; and (c) teaching of the curricular elements and activities to the school groups. Additionally, (d) they participated in activities designed by the museum educational staff to model the hands-on, student-centered philosophy of education espoused by many of the museums and (e) participated in designing and developing pre- and post-visit activities for the school programs that they had been teaching for specific grade levels in multiple curriculum areas. Throughout the three-week practicum, the preservice teachers met both formally and informally with each other and with the museum education staff members to discuss and reflect upon their teaching experiences and learning and development as teachers.

The Outcomes and Impact of Museum Practicum for K–12 Preservice Teachers

The research approach employed to investigate the impact of the museum practicum was very similar to the approach employed in the study of the impact of museum field experience with the MMEd graduate students. Since the inaugural year of the museum practicum in 2005, hundreds of preservice teachers have undertaken field experience in the partner museums. In total, 35 preservice teachers who participated in the program across the years 2005 to 2009 contributed to the database of the reported impact. At the conclusion of their field experience, students were interviewed individually to ascertain the impact of their museum practicum on their conceptions of education and identities as educators. The final interpretation of data resulted in six themes representing the transformation of the preservice teachers' thinking about education, teaching, and learning together with supporting excerpts from the interviews.

Broader views of education and the application of educational theory at play in the museum. The museum practicum offered the preservice teachers the chance to look at education and teaching from a broader perspective. Several of the preservice teachers were of the view that their initial experiences in their bachelor of education degree had been narrowly focused on teaching in the classroom. Preservice teachers appreciated the opportunity to expand their thinking about education to other contexts and to experience teaching and learning in out-of-school settings. They felt the museum practicum challenged them to reflect on the "big picture," and on what is the most important to them in their teaching and on the educational theories espoused in their university courses that were applicable to learning in the museum.

> I find some of the really basic philosophies that we've been learning about are applicable in so many different settings and it's great to see a new setting. . . . I think I came into the education program with a very broad view of what education is, and it became very narrow by just only being in the schools. I found this museum practicum great to broaden again, oh, education occurs in everything we do essentially.

Increased understandings of the educational theory. The preservice teachers frequently asserted that in the museum they were able to see in use, and apply, many of the educational principles that they had learned about from their educational theory classes. For several of the preservice teachers, this was their first opportunity to truly see the learning theory of constructivism in use.

> Having those kids just come in and plunk themselves in front of you when you've never met them before, and all of a sudden, you really

needed to take the time to figure out where they were coming from, what they had done so far, or it was just not going to work. And trying to build on their existing [knowledge], it just became so much more obvious in that setting [the museum]. . . . Yeah, I never got it in a classroom setting [practicum].

Broader skills in teaching students from K–12. The preservice teachers, who were all training as secondary teachers (grades 8 to 12), felt they learned a great deal from their experience of working with and teaching students in all grades. For example, they gained firsthand knowledge about the cognitive and behavioral development of children from kindergarten to high school and learned about what pedagogical strategies work best with the various age groups. They spoke about how the skills they gained in working with elementary children would help them as educators.

I think it was a huge benefit to us to see elementary students. . . . I think all secondary teachers should have experience with elementary because you need to know where they come from. . . . One of the key things I gained was learning a lot about the psychology of kindergarten to grade eleven, especially kindergarten to grade seven because I hadn't ever seen that.

Enhanced skills in flexible pedagogy and increased sense of autonomy to try different pedagogical techniques. The preservice teachers expressed that the dynamic and emergent nature of the museum education setting helped them to develop skills in flexibility in their teaching. In their view, the museum teaching experiences afforded them opportunities to teach in a manner that was responsive to both the interests of visiting school students and their response to the museum exhibits. Furthermore, the preservice teachers felt more freedom to try out different pedagogical techniques in the museum. Because they taught the same museum program several times during the museum practicum, they were able to experiment with their pedagogical strategies and determine what pedagogical techniques worked best.

At the museum you can try the sort of [pedagogical] experimentation that you would really like to do in the classroom, try something else to see if it works, if something failed miserably on the school practicum, I couldn't try it again. But here [in the museum] you can!

Deeper appreciation for the value of working collaboratively. The preservice teachers expressed that the collaborative nature of the museum setting contributed greatly to their professional development as teachers. They were engaged in continuous learning about teaching and education through the opportunities afforded to them in the museum practicum for joint discussion and reflection on their teaching experiences. The preservice teachers valued the opportunity to work closely with one another and with the many museum educators and noted the contrast in the level and nature of collaboration at the museum with what they had seen in the school setting.

> I think another thing that enabled us to do a lot more exploring [of] our own teaching was that we were together in a group; whereas it wasn't just one student teacher, one sponsor teacher, and 180 kids, go! Now [at the museum] we were coming back and reflecting on what had happened in our practicum and what was happening in the museum. There was a real atmosphere that we were really interested in talking about education, exploring how we could become better teachers.

Gains in self-confidence and self-efficacy as teachers. Students gained confidence in their ability to teach and make sound educational judgments over the course of their practicum experiences at the museums. Although these gains varied in their magnitude, there were several notable and profound transformations in self-confidence and self-efficacy. In these cases, the preservice teachers were able to identify clearly that the museum practicum helped them to overcome some of the professional and personal struggles they had experienced in their classroom practicums.

> I have also noticed that I have gained more confidence in my teaching abilities and strategies. . . . I have also noticed a change in my assertiveness with students. Perhaps this is due to the younger age groups that I was dealing with but regardless, this was something that I definitely struggled with in the first part of my practicum. . . . I'm glad to know that I have it in me to be the leader that I need to be as a teacher.

These key outcomes, manifest from the preservice teachers' museum practicums, speak loudly to the limitations and the confined and restricted nature of the school-based practicum alone. The preservice teachers' newly experienced and realized levels of (a) understanding of educational theory, (b) autonomy and emancipation as educators, and (c) self-confidence in their own pedagogical practices again speaks powerfully to the pedagogical and transformative power of museums as sites learning the *art of education*.

CONCLUSIONS

In this chapter, I discuss the value, capacity, and power of museums as places where those who aspire to become educators can learn the *art of education* and the skills to become effective educators. Through the exploration of the cases of two academic programs at the University of British Columbia and their collaborative partnerships with museums in the city of Vancouver, it is evident the museums do in fact have the capacity to serve as highly transformative environments for producing effective and highly skilled educators to the benefit of society.

The case of the museum field experience for graduate students in the masters of museum education program illustrates the considerable power and importance of firsthand experiences in museum settings to develop deep understandings and appreciations of application of theory to practice. It seemed evident that the 120 hours of field experience working in situ with practicing museum education staff led preservice teachers to deeply understand that principles behind museum education can vitally empower visitors' learning processes. Moreover, the personal development as educators through the field experience yielded considerably deep appreciations of how visitors learn and the critical influence of visitor identity on learning. In addition, appreciations of the vital role of the social dimensions of visitor experience and of their personal agendas' influence on their subsequent learning outcomes were major transformations resulting from the field experience. Significantly, the graduate students also came to understand that the nature of learning (and educational practice) in the museum requires an epistemology that focuses not just on the object or artifact and its meaning but also on both the museum as the mediator of content and on the visitor as learner. In other words, focus on the museum object or exhibit alone to the segregation of the visitors in the learning equations is a bankrupt pedagogy and ill-equipped to support rich and meaning learning in museums. This conception together with the notion of museum as an "emancipated" learning environment in which visitors were free to question, talk, and exchange ideas with others contributed to the makings of educators who deeply appreciated museum education as an "art."

The case of the museum practicum for the development of preservice teachers illustrates the power of museum practicum to influence educational philosophy in positive ways. The museum practicum undoubtedly contributed positively to the preservice teachers gaining a holistic view of teaching that went beyond the narrow bounds of the classroom. They not only enhanced their techniques for teaching at the secondary-school level but also acquired appreciation of, and skills in, teaching elementary students. The dynamic and changing nature of the museum education context required the preservice teachers to be highly flexible in their pedagogy and responsive to the learning environment. The museum teaching and learning environment, including the enthusiasm and support of the staff, appeared instrumental in increasing their self-efficacy and positive identities as teachers. Involvement in program delivery and curriculum/resource design at the museums helped the preservice teachers reflect on and identify the value of learning theories they had heard about in their university coursework, but not seen in action. Preservice teachers could see and experience the principles of constructivist learning theory at work—both the individual and social aspects of constructing understanding became apparent in this learning environment. Finally, the preservice teachers "learned about teaching" by working together to discuss and reflect upon their own pedagogy and educational philosophies. This process also helped them to realize the importance and strength of collaboration with others for their professional development in teaching.

Concluding, as is apparent in the two sets of themes, the participating students experienced the process of transforming significantly their philosophy and pedagogy that was being shaped by the museums' educational practices in which they were becoming enculturated. And thus a holistic examination of both case studies clearly demonstrates not only the underappreciated power of museums as places to become educators but also the underappreciated potential of museums to help educators develop the art of being an educator.

NOTE

A similarly themed version of this article was first published in the Japanese language in: D. アンダーソン（2015）「教育実践を 指導し変革する方法を学ぶ場としての博物館の役割」2章湯浅万紀子編 著『ミュージアム・コミュニケーションと教育活動』博物館情報学 シリーズ5巻樹村房. Translated as: Anderson, D. (2015). "The role of the museums as sites for learning how to teach and change educational practices." In M. Yuasa (Ed.), *Museum informatics, Vol.5; Museum and communication, Chap.2*. Tokyo: Jusonbo. The themes of this chapter are reproduced here in English with the permission of the editor, M. Yuasa, and the publishers, Jusonbo.

REFERENCES

Anderson, D., Kisiel, J., & Storksdieck, M. (2006). Understanding teachers' perspectives on field trips: Discovering common ground in three countries. *Curator, 49,* 365–386.

Anderson, D., Lawson, B., & Mayer-Smith, J. (2006). The impact of extended practicum experiences in a marine science centre. *Teaching Education, 17,* 341–353.

Hein, G. (2005). The role of museums in society: Education and social action. *Curator, 48,* 357–363.

14 Breaking Dichotomies

Learning to Be a Teacher of Science in Formal and Informal Settings

Preeti Gupta, Cristina Trowbridge &
Maritza Macdonald

This past decade has yielded a deeper understanding of how formal and informal learning contexts are not at odds with one another but rather complement each other. The formal and the informal environments create ecology of learning spaces. When learning occurs at the nexus of formal and informal learning spaces, leveraging the affordances of both spaces, that is when students not only engage with the endeavor of learning science but also are prepared to carry out the practices of science. Teachers need to be prepared to maintain an identity of not just a schoolteacher, but a teacher of science, one that crosses boundaries of teaching in both formal and informal spaces and one that espouses a reform-minded identity steeped in a world-view that privileges science inquiry, where the teacher believes that a learners' prior knowledge and everyday experiences are critical in the process of learning, the goal being to steer youth toward science literacy by deepening understanding of specific core foundational topics (Avraamidou, 2014).

Informal science learning environments can be broadly defined as non-school settings that offer opportunities to experience and make meaning of scientific phenomenon. These environments range from a schoolyard to an amusement park. For the purpose of this chapter, informal science learning spaces are structured institutions, such as museums, zoos, aquaria, and gardens. These institutions are where people can engage with science in ways that are authentic and participatory. The resources in informal science learning spaces include exhibits, collections (living and nonliving), educators, scientists, and science labs. These resources are important in supporting teachers and, in particular, those learning to become teachers, because they afford the opportunity to enact culture in particular ways that contribute to reform-minded teaching (Luehmann, 2007). The affordances of learning to teach in such settings include opportunities to work with multiple and diverse audiences thereby mastering one's skill at student engagement, being comfortable to teach the same content to different audiences, time to refine the lesson, and the opportunity to develop an awareness of self as a teacher and a learner, which all contribute to identity development as a science teacher (Gupta & Adams, 2012). We take teacher identity to be "the ways in which a teacher represents her views, orientations, attitudes, emotions, understandings, and knowledge and beliefs about science teaching and learning" (Avraamidou, 2014, p. 826). We use this definition to illuminate how a person's identity as teacher of science becomes visible through enactment of culture and is always transforming. The practices of that teacher are an embodiment of her ideas and beliefs and how she uses available resources.

In this chapter, we examine the affordances of museum resources in informal settings and how they shape science teacher identity. More specifically, we discuss how residents learning how to teach in school settings leverage experiences of learning to teach in museum settings. We describe a masters of art (MAT) program that is situated at the American Museum of Natural History (AMNH). The MAT, an earth science teacher preparation program, is a 15-month residency program. People with undergraduate degrees in earth science/astronomy or a closely related topic can apply with the intention of becoming certified to teach earth science in New York State. These people hereby referred to as residents, undergo a program that begins in June and consists of two museum residencies and two school residencies. Residents also complete 36 credits of science and education courses, which are co-taught by science and education faculty, many from AMNH. The two museum residencies serve as bookends for the 15-month program. Upon graduation from the program, the cohort of teachers participates in a two-year new teacher induction program at AMNH.

In the first museum residency, teacher candidates rotate through three distinct museum activities that build on their instructional abilities to use informal science resources. This eight-week residency provides experiences with everyday museum visitors and middle and high school students enrolled in museum youth programs. In the first two weeks of the rotation, residents work in small teams with interactive museum touch carts in the museum halls. Residents spend about 20 hours with the museum touch carts in the halls, which provide many opportunities to engage and access visitors' prior knowledge, and introduce science concepts and assess for understanding. During the second rotation, residents get an in-depth experience observing and co-teaching with museum educators while working with middle and high school students who participate in weeklong and month-long museum youth programs. In the third rotation, the residents have the opportunity to design and teach their own lessons for high school students who are invited for a summer science institute. During the museum residency, residents take a course that provides the theoretical foundations for learning in informal settings.

In the second museum residency, residents spend time learning in the field and doing research with museum scientists. This experience allows residents to collect rock and mineral specimens to use in their instruction. This chapter focuses on what we are learning from the first museum residency and how teachers are enacting museum resources in their own teaching once they become teachers.

MAT graduates participate in the new teacher induction program, which takes a multilevel approach to supporting and developing new science teachers. The program consists of school visits with classroom coaching, monthly meetings at the museum, and 30 hours of professional development. During this time, teachers become part of a professional learning community, share their student work as well as participate in activities to improve their facilitation of student discussions and deepen science conceptual understanding using museum resources. To date, 50 candidates from three different cohorts have graduated and are teaching in high-needs public schools (defined as schools with at least 70% of students living below poverty level). The cohorts consist of 50% men and 50% women, 18% Black or Hispanic, and 37% career changers.

EMERGING IDENTITIES AS TEACHERS OF SCIENCE

Learning to teach is a cultural activity, one that is shaped by the contexts of the environment but also shaped by one's own sociocultural and political standpoints. We describe cultural activity as occurring in a field, where culture is produced, reproduced, and transformed (Sewell, 1999). Fields are sites of activity with porous boundaries and have structures, both visible and invisible and those structures consist of resources (material, societal, cultural) and schema. Resources may include objects, people, tools, and rules, and schema are the ideals and beliefs one holds. One's practices are a set of actions that draw from one's schema and available resources. If we claim that the museum is a field, then we can say that as residents work in these fields, they are using the structures of the field to afford them the power to act, to have agency. As a resident becomes agential in the field, she actually mediates changes to the structures of that field. One way to visualize is by using the Scheffer stroke "|," which denotes dialectical relationships (Roth & Lee, 2007). Then we can represent the concept as such, structure | agency where structure is defined by the dialectical relationship of resources | schema.

When residents operate museum touch carts, they are enacting culture in the museum. They are using the resources and schema available to them. At each interaction with a visitor at a cart that contains objects, the resident has to act in compliance with the resources available and the schema that exists in that moment in time. The person also has to experience passivity (Roth, 2007). After a resident asks a question or poses a challenge, the visitor's actions and words are unpredictable. The resident has to accept this reality and experience this state of being and then be ready to act in response to the moment before, and use the structures available in that new moment. In this way, each interaction becomes a time to use and simultaneously transform structures. Each successful and unsuccessful interaction mediates the way resources are used and schema is developed. After 20 hours of facilitating with the museum touch carts, their agency at teaching at the carts is greater than it was when they started. Because it is a dialectical relationship, as their agency increases, the structures also change at that cart. The way the objects are used and the nature of the conversations with the visitors are different. The resident develops the ability to act and maneuver in different ways in response to visitors. She develops strategies for how to engage visitors. In the vignette that follows, we see how one particular resident discovers the value of creating a challenge at the museum touch cart.

> During my first rotation at the Ocean Life museum cart, I had two children extremely excited about finding out about the sperm whale's tooth. Since they were so hyper I sent them down to see the diorama and told them to come back to tell me what they have seen. A few minutes later, they came back but they didn't have the right answer so I asked them if they wanted me to tell them or they wanted to continue their adventure. They decided to go back downstairs to check again. When they came back they were jumping and screaming that it was a sperm whale and a huge squid. But when I asked them to whom the tooth belonged to, the boy said it was a squid's tooth and the girl said that it was the whales. I had the boy tell me why he told the tooth belonged to the squid and then I had the girl explain to the boy why the tooth belonged to the whale.

Natural history museums are built on the foundation that objects in their collections tell an important and compelling story. Learning how to teach with objects takes time, and one needs to develop strategies that are comfortable, effective, and make sense for the curricular goals. In the vignette, the resident has discovered the power of engaging learners by posing challenges and reveals how she connects an object to a set of exhibits. In the following vignette, we see how a different resident is developing an appreciation for object-based learning but also incorporating a learner-centered pedagogical approach.

> When a large group gathers around the cart, it takes much more awareness on the part of the educator to keep everyone involved and engaged. I am far from mastering this skill, but so far have found that passing the cart objects around the group and getting everyone involved in brainstorming answers to a more difficult question help maintain the large group interactions. At this point in the museum residency I am just beginning to explore what techniques and resources I find most effective in certain context (e.g. different age groups, cultures, group sizes). I have walked away from this past week's discussions, readings and observation of cart use with a better understanding of the importance of object-based learning and the role that questions can play in creating a more engaging and open learning experience.

The two vignettes illustrate the trajectory of how the residents develop a comfort with using the objects by trying different strategies with different people. Short interactions with numerous visitors' means that each time the resident uses the resources of the museum touch cart and engages the learner, it mediates changes in her schema. Each interaction makes her think about learners and learning differently. As her ways of thinking about how to interact with people changes, her use of resources on the cart changes. As the resources | schema dialectic continues in each interaction, her practices change. As the residents are successful with visitor interactions, they are seen as a certain kind of a person by their peers and by visitors, a person who knows how to teach science. Over time, they begin to see themselves as successful teachers of science, and this mediates changes in their identities. Identity development becomes intertwined with the act of doing. Because the construct of identity is dynamic and fragile, often, unsuccessful interactions can lead to threats to one's ability to teach. However, because residents are required to spend 20 hours on the museum touch carts, there is the opportunity to learn from the unsuccessful interactions and to use resources and apply practices differently. Unsuccessful interactions can become a way to revise a teaching strategy, which then can lead to a successful interaction. The unsuccessful interactions, which may emotionally challenge one's self-perception of being able to teach, are followed by a successful interaction, which balances emotions. Because the nature of working at museum touch carts means many short interactions in a chain, each interaction becomes a chance to start over, and the rate of successful interactions is far greater than the rate of unsuccessful ones. A resident begins to identify with what teaching with objects means and this is not a shallow understanding, but one steeped in multiple lived experiences, with each time being given a chance to apply practices differently as governed by the resources | schema dialectic.

THE AFFORDANCES OF EXHIBITS

Whereas objects can be brought into another learning space, exhibits often cannot be replicated in the same way. Field trips become the primary way for accessing and using the exhibits for schoolteachers. Field trips, especially for new teachers, can be a daunting task for reasons that include uncertainty of the teacher's role, lack of comfort with the content, logistical challenges, and administrative obstacles (Kisiel, 2014). To combat some of these issues, the MAT program aims to develop comfort and confidence at facilitating field trips. The second rotation in the museum residency places residents in co-teaching roles with various expert museum educators who lead a three-week summer program for middle and high school youth who are in a museum program that is designed to teach key concepts of anthropology, biology, and physical science in ways that fully integrate the exhibits of the museum with classroom activities. The summer program is designed so that students visit exhibits each day and over the three-week period with different museum educators. When residents are working with a number of the museum educators throughout these three weeks, they observe multiple teaching styles and multiple approaches to using the exhibits in a lesson. In the following vignette, we experience a resident who believes in the value of informal learning spaces and is pondering how to use non-classroom experiences to augment the formal learning space.

> The instructor for the sixth grade biodiversity program jumped into museum on the first day with visiting the Halls of Biodiversity and Ocean Life. He then followed up with another assignment that required a visit to the Hall of Reptiles and Amphibians. The kids also did an activity that required them to virtually gather anthropological items to create their own museum collection.

Having multiple opportunities to see how children get engaged at exhibits, get excited, and bring that excitement into the classroom is important for aspiring teachers, because it allows them to imagine and understand what good museum learning experiences can look like. When residents experience and compare the different pedagogical approaches taught by the museum educators and observe the way that children respond to those approaches, they are able to create a schema for themselves that consists of strategies and successful experiences. The museum educators have their unique style, but all use instructional strategies to engage children with objects on display in a natural history setting. During this rotation, residents observe and reflect on how particular approaches are used, such as sketching dioramas, talking in front of large-scale specimens, and using worksheets to gather evidence in exhibits. The residents realize that using the exhibits as a resource and with a particular schema associated with it allows children to exhibit their interest and abilities in multiple ways.

Residents take courses throughout the academic year that follows the first museum residency. In one of the courses, there is an assignment to develop, implement, and assess a museum investigation that is embedded in a specific unit of instruction for students at their residency schools. The goal of the assignment is for residents to build on their summer museum residency experience by integrating museum resources into the school curriculum. It is the nexus of bridging the formal school with the informal learning experience. Residents are asked to reflect on student work and share insights for the evidence for student learning. When the residents conducted their museum investigation assignments in this course, it was not surprising to see that they chose numerous exhibition halls within the museum where they had spent time during the museum residency. Residents were asked to reflect on the field trip experience and in particular focus on the work produced by the students on the trip. In the vignette that follows, we see how students in a special education class thrive during the museum experience. The resident states,

> I took seven kids from a marine science class which I have been taking the lead on for most the semester. Attendance and excitement was just not there. They attended irregularly. They needed something. Excitement about the subject was there but museum trip would be perfect. It would get them excited about anything they wanted, but I would try and direct them to the ocean life stuff when we got there, because that is what we were doing. We had done an intro before, to get a sense of marine environment. I tried to do a lesson but it was so boring for them even though we had some visuals, but they weren't great. But then, dioramas on top of the Ocean Life Hall you can see the environment, they are everything short of interactive. They loved it. Then I said pick your favorite. They all went to the polar seas one. They got excited about that part of it. After they were done with their worksheets. It was nice to see them put down the sheets and actually look, instead of looking with something in mind, and then they had all kinds of questions and their work was fine, it was accurate.

Because the visit was a field trip as part of a formal school structure, worksheets and assignments were used to check for student understanding. The aforementioned resident had a reform-minded approach, one that was learner-centered, and grounded in the learner's interests. For the assignment, she asked the students to choose an ecosystem to learn about it in the Hall of Ocean Life. She leveraged the free-choice learning mantra of informal learning settings to give them power over their own learning. A number of the residents reported that because of the experience in the museum residency they knew the value of letting their students have options for their own learning on the field trip. For this to happen, residents needed to trust the students and trust the potential outcomes of having them lead a portion of their own learning experience.

The museum investigation provided an opportunity for this resident to get to know students in a different way. The museum and the school are both fields of activity, but these fields have porous boundaries. Residents who have learned to use the structures of the museum carry that agency and associated structures into their formal classroom space and then back. As they enter back into the museum learning space, they now have the structures learned from a formal environment mediating their agency. The dichotomy of formal and informal learning spaces breaks down. The structures and associated agency developed in both fields now mediate activity in both fields.

THE JOURNEY CONTINUES

The first three cohorts of the MAT residents are now first- and second-year middle and/or high schoolteachers. The teacher shifts from being a student in the MAT program to teachers in new teacher induction. We observe the shift in their identity from MAT resident to new science teacher. One of the goals of induction is to use the affordances of the museum to nurture the teacher's own informal learning and strengthen their ability to use museum resources, especially while their new teacher identity is influx. The shift of forming identities as reform-minded teachers takes time and is influenced by context and experiences early on in teaching. One activity in the induction program is to support teachers in developing their abilities and nurturing their own passion for learning. Teachers engage in various strategies to observe, draw, and talk about a specific diorama or museum exhibit. Afterward, teachers reflect on how to translate these types of experiences into their classroom instruction.

By the time the teachers begin new teacher induction, they are comfortable planning museum learning experiences and using objects, but whether they actually plan and carry out depends both on their school context and their identity as a teacher and learner. The MAT teachers exhibit practices that convey their interest in using objects and the physical resources at the museum. In the first four months of teaching, 56% of the teachers ($n = 42$) made a visit to AMNH and by the end of the first year, 85% had visited AMNH or another informal learning site. The teachers who visited late in the year regretted not taking their students earlier because of how much the students learned and how the teacher was comfortable facilitating in the museum.

In documenting the experiences of particular MAT graduates who brought students for a field trip, we saw evidence of their transforming identities as people who feel the museum is an extension of their classroom and is a place where they can engage their students. We also saw the power of being seen as a certain kind of person by both students and administration, a person with access to the museum, and a person with passion and expertise in the science.

Dan

One teacher, Dan, was struggling with making his lessons active and was met with some resistance by his students. He was challenged to connect with his students and was afraid to take a field trip. He did so after hearing about another MAT graduate who taught in a similar school setting and how the field trip was successful for the students. In the third month of school, Dan brought his class to the museum, and he began to use the museum resources such as the geology exhibits that he knew well and drew upon his associated schema for teaching with those resources. He exhibited confidence, a comfort, and an agency in teaching his students. He later reported that the visit became a turning point for his relationship with his students. Dan had a museum badge, access to particular exhibits that normally cost additional money, and was being treated by the museum as if he belonged there. He stated that students talked about the trip once they returned to school, and he received some "cred" in the student's eyes for providing a trip that did not cost money and had extra access. They also respected his ability to use the exhibits with ease and passion. Dan's practices were comfortable in the museum because he had the agency to use the exhibits proficiently. His students saw this fluent practice and now identified him as someone who knows science, is passionate about science, and is linked to an institution of science. Although he had the ability to practice reform-minded teaching in the school environment, his students were not affording him the structures necessary for success. When they experienced his practice in a field that was outside of their comfort space, their schema about what was possible changed. When he returned to a formal school structure, the students were carrying back this schema and afforded him the space to teach science and to use the agentic practices he exhibited at the museum.

Mina

For some teachers, whereas field trips were not options immediately available because of school structures, their identities as reform-minded teachers allowed them to exhibit themselves in different ways. Mina's story provides an instructive example of bringing the museum resources into the classroom. Mina placed pictures of herself in the field collecting rock samples with her MAT colleagues on her classroom walls. She had rock samples that she had collected showcased in her classroom. One of her first student assignments was to have students take their own photos in a place where geology is found in the city. In her classroom, she had many of these photos of her students standing in a park, near a river, or on a rock. Throughout the year, Mina kept using museum resources in her classroom. She invited two scientists to her class that she had worked with in the MAT program. Although it appeared that she is an ideal candidate who would conduct a field trip, she was challenged by school administration.

In February of her first year of teaching, she asked her principal if she would be able to bring students and parents on a weekend day to visit AMNH. She came to realize that if she visited on the weekend, her principal really could not refuse. She worked with the parent coordinator and planned a trip for a Sunday in May for about 70 students and 50 parents. The trip was pivotal at stabilizing her teacher identity. One of the parents who went on the trip was so impressed with how much the students and parents learned, the parent called to inform the principal. It was this call from a parent that shifted the principal's idea toward trips. But it did something for Mina also. Mina believes her principal saw her as someone who had "cache," a resource by having an affiliation with the museum. The principal's ambivalence toward trips dissipated, and Mina stepped into taking a leadership role at the school leading to a number of school partnerships with AMNH. Mina had the agency to keep bringing resources into her classroom and ultimately planned a weekend trip. Being seen as a certain kind of person by the parent mediated the actions of that parent to talk with the principal, which led to the principal to see Mina as a certain kind of teacher—a teacher with access to AMNH resources and, more critically, the ability to engage students and parents using the affordances of the museum.

IMPLICATIONS

Informal science education institutions are critical partners for teacher preparation. The opportunity for aspiring teachers to use the affordances of such settings allows for a shaping of identity that blurs the lines of formal and informal learning and allows teachers to use the resources of such settings in varied ways to engage learners, garner respect for themselves from both adults and students, and develop a sense of self as a teacher of science that is strong. We describe a number of affordances that museums offer that position aspiring teachers to practice teaching in low-stakes settings, with objects, to diverse learners and work alongside many different experts, each time building a framework for what good teaching looks like. Throughout the program, opportunities to exercise those strategies are present, leading to relative comfort and familiarity with conducting field trips. As these aspiring teachers become teachers of record, we see threads of evidence of how the particular schema trigger practices that lead to using the affordances either in the form of science activities, object-based teaching, or science museum investigations.

There is much to be learned as the AMNH MAT continues to graduate more science teachers. In particular, we are curious about the extent to which MAT graduates are using informal resources in instruction to strengthen student understanding and engagement in science. However, we hope that the ideas presented will support others to document their experiences working with teachers in informal settings and to build on how these experiences shape teacher identity and strengthen instruction for all students. There are more than a thousand different science museums in North America and many of them are situated near institutions of higher learning. Many already have partnerships with each other. What are the opportunities to develop ways for aspiring teachers to work in such places where they can practice teaching in low-stakes settings using different exhibits and objects with intergenerational visitors? How can museum faculty work with higher education faculty to create structures and processes to weave theory into practice? How can museums be used for supporting new teachers through induction experiences? These questions and many more are critical for consideration as we experiment with teacher preparation at the nexus of formal and informal learning settings.

REFERENCES

Avraamidou, L. (2014). Developing a reform-minded science teaching identity: The role of informal science environments. *Journal of Science Teacher Education, 23,* 823–843

Gupta, P., & Adams, J. (2012). Museum-university partnerships for preservice science education. In B. J. Fraser, K. Tobin, & C. McRobbie(Eds.), *Second international handbook of science education* (pp. 1147–1162). New York, NY: Springer.

Kisiel, J. (2014). Clarifying the complexities of school-museum interactions: Perspective from two communities. *Journal of Research in Science Teaching, 51,* 342–367.

Luehmann, A. (2007). Identity development as a lens to science teacher preparation. *Science Education, 91,* 822–839.

Roth, W.-M. (2007). Theorizing passivity. *Cultural Studies of Science Education, 2,* 1–8.

Roth, W.-M., & Lee, Y.-J. (2007). "Vygotsky's neglected legacy": Cultural historical activity theory. *Review of Educational Research, 77,* 186–232.

Sewell, W. H. (1999). The concept(s) of culture. In V. E. Bonnell & L. Hunt (Eds.), *Beyond the cultural turn: New directions in the study of society and culture* (pp. 35–61). Berkeley, CA: University of California Press.

15 City-as-Lab Approach for Urban STEM Teacher Learning and Teaching

*Jennifer Adams, Eleanor Miele &
Wayne Powell*

Urban settings are great places for science teaching and learning because of the range of science-rich places available. Ranging from the streets of the local community to brick-and-mortar institutions, there are a myriad of places with the potential to engage teachers and their students in science learning. Leveraging these resources, we developed the City-as-Lab approach to urban place-based science teaching to prepare teachers to integrate these resources into their teaching, extending science learning beyond the classroom. Over the years and through our practice, we have grown to recognize the salient connections between place-based education and teacher identity development. Through collaborations with informal science institutions and the integration of faculty field-based research in science teacher education, we connect teachers to local places for science learning. In this chapter, we first describe the rationale and theoretical underpinnings of this approach. We then elaborate on the City-as-Lab framework to illustrate the relevance of place-based teacher experiences in developing teacher identity, concluding with anecdotes and evidence on the impact of City-as-Lab.

TEACHING AND LEARNING TO TEACH SCIENCE IN AN URBAN SETTING

Our college is situated in an urban context, and as such we have been able to use our professional expertise alongside our personal lived experiences and identities as educators to shape a teacher education program that is both meaningful and relevant to urban students. Central to our thinking around the design of earth science teacher education programs are the diversity and character of the people who live in our urban setting and the physical settings that most engage students. We think of an urban context as a place that (a) is culturally, ethnically, and economically diverse; (b) has a landscape that is dominated by a built environment; and (c) is often the cultural and economical center of a region. This means that although there are challenges to teaching and learning in urban contexts, there is also vast opportunity in the range of resources, including places and spaces available to teach and learn science in ways that both parallel and complement standards-based curricula. In addition to the range of physical places, other tools of science such as historical, digital, and GIS resources help learners to view their cities as dynamic, interconnected systems and allow them to explore local ecology, infrastructure, and geography in relation to their lived experiences. We use this physical and social context to teach teachers how to leverage their place—both in their identity as teachers and as situated in an urban context—to create curriculum that allows urban students, especially immigrant, racialized, and economically marginalized students, to participate in rich science learning experiences.

Urban students face a myriad of systemic challenges to success in science, including, but not limited to, having access to inadequately resourced schools, often staffed with minimally qualified science teachers (Powell, 2005). Along with the intersection of race and ethnicity, Black or African American, Latina/o, and immigrant students are often marginalized from educational experiences that afford equitable access to and successful outcomes in higher education, including science pursuits. However, research has established a more positive picture of science learning in informal learning environments, especially for these young people. Informal learning environments allow youth to interact with science in ways that are personal, socially meaningful, and relevant to how they view science in their lifeworlds and develop positive identities and agency, both in and out of school, in relation to science (Adams & Gupta, 2013). These researchers, and others, advocate that teachers create classrooms where students are able to integrate the community-based and relevant knowledge that students bring with them into the classroom with their science learning experiences. Thus teaching teachers how to integrate informal science learning, with an emphasis on place, in the formal classroom will expand learning opportunities for their students and allow for a more equitable science teaching and learning space to emerge.

PLACE, IDENTITY, AND LEARNING TO TEACH SCIENCE

In taking a place-based approach to learning, we not only consider the physical setting but also the institutions, cultural meanings, human interactions and possibilities that are associated with places. Although we are concerned with the science connected to place-based resources, we also recognize the multiple layers of meanings people ascribe to places and how this influences place-based learning (van Eijck & Roth, 2010). This resonates with our description of learning as an enactment of lived experience, inextricably linked to place, as this is how both teachers and students make meaning of their worlds and accordingly develop identities and associated practices.

In order to make salient connections to places in the context of teacher training, it is important to consider the process of teacher identity development. A definition of teacher identity that resonates with our work is one that describes how a teacher represents herself to others through her enactment of teaching (Avraamidou, 2014). Teachers represent who they are as educators through the practices they choose to enact, including the places they select to incorporate into teaching, the communities with which they align, and the narratives they tell about their teaching.

Teacher identity development in a place-based framework is interconnected with place-identity development. We take place identity to be the relationship between a person's identity and the physical environment, complicated by a host of factors, including ideas, values, goals, skills, and affect relevant toward a given environment. Gross and Hochberg (2014) describe four indicators of place identity: familiarity, involvement, belonging, and meaningfulness. They found that nurturing these characteristics positively affects the development of other components of identity connected to science teaching. In our work, we aim to connect places through these factors as associated with science teaching and learning, while not dismissing their existing place identity relationships. Through teacher education, we hope to add a science-rich layer to teacher place identity that they will then share with their students. Developing connection with places as resources for science teaching is salient in developing practices and a corresponding identity that ameliorates the boundaries between the classroom and the local community. We believe that it is important for future urban teachers to make these connections between science and students' lived experiences.

STRENGTHENING TEACHERS STRENGTHENS FUTURE STEM MAJORS

As a local public college interested in recruiting strong, local students, we had an incentive to prepare local teachers with the knowledge, skills, and dispositions that we believed would contribute to their students' success in science coursework at the college level. It is important for us to both build a strong cadre of earth science teachers and a strong pool of potential majors for our earth and environmental science and teacher education programs. In our equation, strong science teachers equal strong science majors. Because much of our research focuses on the local environment, in both earth and environmental science and in science teaching and learning, the notion of developing teachers and local students who share our commitment to place is also important to us.

OUR TEACHER IDENTITIES INFORMED OUR APPROACH

Our own identities, as faculty members, were relevant to creating and maintaining the City-as-Lab approach to science teaching, learning, and learning to teach. We (Jennifer and Eleanor) were born and raised in Brooklyn, where our college is located and have fond memories of visiting the range of cultural and natural resources available. One of us (Eleanor) had spent many hours as a disaffected high school student roaming the halls of the natural history museum in search of meaningful learning. These visits carried over into her emphasis on observations, inquiry, and collecting authentic data in the experiences she creates for teachers and her understanding of the value of place to motivate learning. Jennifer fondly recalls family and class visits to the same museum. As a high school biology teacher, she lead class trips to the museum, even with administrative roadblocks, because she knew from her lived experiences that the museum was an inspiring science place. Years of teaching and research in formal and informal contexts taught her how to enact meaningful and to effect learning experiences for science teachers and students. Wayne's field research and authoring of field guides fueled his desire to connect students to place-based science learning. In reflecting on our own science learning, while we learned in schools, we always identified strongly with the power of place to fuel our interests.

The Urban Context as a Science-Rich Lab

Our City-as-Lab approach centers around three key principles: (a) a thematic approach to integrating community-based resources, (b) authentic science experiences, and (c) collaborative team-teaching and peer-learning. Our goal is to educate teachers who would use this approach to integrate the resources of the city into their science teaching, therefore, it is both our approach for program design and teaching and our hope for pedagogical enactment in urban classrooms. Learning experiences are designed so that graduates of our programs will be able to:

- integrate New York natural and cultural resources into their earth science classrooms,
- recognize the connections between earth sciences and the lives of the people of New York, and
- be part of a supportive and cooperative network of educators (including both teachers and informal education professionals) and geoscientists.

The City-as-Lab approach began with the redesign of the initial earth science teacher certification program and continues as a framework for planning grant-supported programs and professional development for teachers. What follows is a discussion of the program design process illustrated by an NSF-funded project that connected science teachers with faculty around place-based science research.

THEMATIC APPROACH TO INTEGRATING
COMMUNITY-BASED RESOURCES

Place-based learning allows teachers to develop attachment to places (place identity) through both direct experiences and instructional activities

(Kudryavtsev, Stedman, & Krasny, 2012), with the instructional activities ideally modeling effective pedagogical approaches. In our planning of the City-as-Lab, we wanted our teacher candidates to have direct experience with places—through scientific data collection or, in the case of museums, interactions with museum exhibits for science learning. Because we are also interested in our teacher candidates successfully completing certification requirements, we structured our content goals around both these require-ments and the local high school exam standards. However, our pedagogi-cal goals were focused on learning content in context; extending the earth science classroom to the streets surrounding the school and beyond. In our planning of the program, we consistently revisited our guiding question, "In what ways is this a uniquely place-based theme?" This allowed us to keep the focus on place central to our planning and focus our pedagogical discussions around teaching teachers to access and appropriate the resource of place in their teaching.

Our design process mirrored the collaborative approach that is used in all courses in the program: four groups composed of at least one team mem-ber from each stakeholder group—classroom teachers, faculty members, and park or museum staff—were tasked with the goal of creating practice-related understandings of City-as-Lab. All participants had a place-based practice of teaching and/or research and therefore brought and shared dif-ferent perspectives and professional identities around the common theme of place-based science teaching and learning. Each team was asked to identify thematic questions that could be productively explored using resources in the city and realistically enacted in the classroom within the constraints of time and curricula. At the end of several brainstorming sessions, we identified five thematic courses and essential questions that together would address all required content areas and essential skills required for teaching earth science to local students (Table 15.1).

The resulting themes (Table 15.1) not only emphasized particular places that define our city but also addressed questions that are relevant to teach-ers and students understanding the interconnection of geological processes and the built urban environment. For example, "New York City Water" allowed participants to explore the infrastructure of an everyday and often taken-for-granted resource. Themes such as "Geology in World Cultures" and "Planetary Science" allow students to access cultural and science muse-ums to develop essential understandings beyond the local through the global artifacts on display, which are also available to students and teachers every-where through online collections and resources. Through learning science and learning to teach in and with different places, teachers develop identities that allow them to view their place differently. One teacher noted that after taking the Geology in World Cultures course, he never looks at a building in the same way. In a city that is dominated by stones fashioned into structures for living and working, being empowered to infer the origin of stones, finding fossils in building lobbies, and connecting building structures to underlying

Table 15.1   Themes and essential questions, level required, and partner institutions.

Theme and Essential Questions	Level	Partner Institutions
Geology in World Cultures: How do the arts and artifacts of global civilizations reflect global geology?	Introductory	Metropolitan Museum of Art; Brooklyn Museum
New York City Water: Where does it come from and where does it go?	Introductory	NYC DEP Treatment Plants; Local Reservoirs; Prospect Park waterways
Planetary Science: What is a planet? What are the properties of objects found in the solar system and beyond?	Introductory	AMNH Rose Center for Earth and Space
New York State Geology: How does the geology of New York State reflect the major concepts of plate tectonics?	Advanced	Museum of the Earth, Ithaca, NY; Multiple field-based sites in Upstate New York
Global Cataclysms: What is the geological evidence of global cataclysmic events that have impacted the evolution of life on Earth?	Advanced	National Parks of New York Harbor National Seashore

geographic features provided him a meaningful way to connect with his city as a science teacher. He also mentioned that his students now look for fossils everywhere—their science-place lenses have also been expanded.

AUTHENTIC SCIENCE EXPERIENCES

For us, authentic science learning means connecting learners to places through data collection and analysis that address real scientific questions that range from faculty research to community-based environmental issues. The thematic courses embed experiences that empower teachers to use the tools of science to collect data that address research or local environmental issues. Many of our teachers who graduated with the City-as-Lab framework continued to be informally involved in place-based science research and to participate regularly in informal science learning professional development opportunities. This indicates that many of our teachers developed professional identities around the resources that they were exposed to during their teacher education. We received an NSF-funded Geoscience Research Experience for Teachers (Geo RET) supplement with the overarching goal of advancing earth science beyond mastery of geoscience content and pedagogy (with an emphasis on preparing teachers to empower students to collect and analyze data at the middle and high school levels) and toward mastery of professional standards for geoscience inquiry. Through the Geo RET, we institutionalized these connections and provided credit-bearing options both for new and experienced teachers that provide opportunities for them to confirm and deepen these salient place-/practice-related identities. During a yearlong cycle, teachers received six course credits for participating in a summer research immersion and completing an independent project as a member of a research team connected with a faculty mentor's lab. Faculty mentors' research included urban tree pit water management, coastal marine debris removal, and examining the local fossil record—all local research projects with two focused on local urban environmental sustainability.

We conducted an evaluation of the teachers' participation and learned key lessons about the relationship between the experience and their teaching identity. Whereas we did not assess the place-identity component, we are able to make conjectures about this experience and its connection to place through subsequent informal discussions and exchanges. First, learning about the nature of doing scientific research was prominent. One participant noted, "being out in the marsh taking samples . . . select[ing] an appropriate site for study, parsing a collection of scholarly articles to establish our protocols and being out at our site monitoring results have been truly inspirational." For this teacher, the place component was a compelling part of her experience. The place of being allowed this teacher to "*know as*" a scientist rather than "*knowing about*" science, which Tobin, Roth, and Zimmerman (2001) describe as being ontologically different terms. "Knowing as" is taking on an identity, if even for a moment, as someone who is able to do. In the physical setting, this teacher enacted data collection under the conditions of working as a field scientist, an identity that she carried to the classroom through pictures, stories from the field, specimen samples, and, important to this discussion, the recreation of a similar field experience for her students during the following semester. The experience also carried over to the classroom through changed teaching identities:

- "I feel that I have gained more credibility in the classroom in that I can tell my students that I have made new scientific discoveries [discovering a new species in the field]. I am not just teaching the discoveries of others; I have contributed to our knowledge of science myself."
- "I feel that the GEO RET bought me a lot of credibility with my students that I am well-versed in science and science practices. I keep my poster [from the GSA conference] displayed in the classroom and I share my GEO RET experiences with them during an orientation at the beginning of the year."
- "Since I teach science research it is very valuable to be able to share my own personal experiences with students. It gives me more credibility in their eyes that I am not just a science teacher but a scientist conducting

my own research. I hope to continue to participate in research projects
so that I can maintain this connection with my students."

In these three instances, the teachers' participation in the research was important to their "street credibility" as scientists in the classroom. The research
provided experiences that they could share with their students that often
started with, "When I was in the field . . . ." The science research embedded
in place allowed them to connect with their students in a more meaningful way. All of the teachers wished out loud for their students to have the
same experiences, citing the engagement and inspiration they felt from doing
place-based research. They learned more about local environmental issues
and were able to connect those with their classroom teaching. When asked
about her continuing to do geoscience research, one teacher responded, "Yes!
Through my curriculum and place-based learning projects in areas such as
Jamaica Bay and Prospect Park," two local places that people frequent for
recreation. These have now become places for this teacher to enact a science-
related place identity. Other teachers conducted field trips to their research
sites and continued to seek out ways to maintain their own professional
learning connected with place-based science teaching and learning.

COLLABORATIVE TEAM-TEACHING AND PEER-LEARNING

Learning through co-teaching. Science teachers learn and enact who they
are as science teachers through the communities with which they form
alliances and through the practices they accept (and reject). They position
themselves as teachers aligned with particular approaches to science teaching and learning and through this positioning they represent themselves to
others associated with their professional role. Faculty also have teaching
identities, and the communities with which faculty form alliances are equally
important in their identity development as educators. For us, this revelation
was important in developing a collaborative team-teaching/peer-learning
model within the City-as-Lab that would facilitate the exchange and growth
of practices between classroom teachers and science faculty. The teacher
education courses were designed as team-teaching experiences that pair
an earth and environmental science research faculty with an experienced
classroom teacher who had graduated from our program and maintained
a practice of integrating place-based resources and learning experiences in
their classroom teaching. Collaborative team-teaching allows for the shared
experiences necessary in developing an integrated practice around authentic
science and place-based learning. The teachers deepen their content knowledge and understanding about the nature of scientific research, while the faculty both improve their pedagogical practices and expand their knowledge
about the realities of science teaching in urban classrooms. One example of
a mutually beneficial outcome of an enacted faculty-teacher alliance was the
development of a local urban hydrology citizen science project created in

response to discovering a shared need: teachers' desires for authentic data collection experiences and a faculty member with an ongoing need for local data collection. This faculty member collaborated with classroom teachers to develop data collection experiences that are meaningful for students and teachers and are realistic for a daylong field trip to a collection site. The faculty member benefits by having students provide valid data that he can utilize in is local research. As a researcher, he has also become more concerned with public scholarship and the community-based relevance of his research and has adjusted his ongoing research as such. In addition, he and other geoscience faculty have become more involved in research in geoscience pedagogy and even self-identify as geoscience educators, which have impacted their practice in all earth and environmental science coursework.

For teachers, the co-teaching relationship provided new insights about both their own professional learning and the learning of others.

> I'm teaching, but I am also learning about how they [students of science education] are internalizing what I say. I can relate that to when I am teaching my students—if an adult does not understand what I am saying, then a 14-year-old will have more trouble.

In this collaborative environment, the teacher was able to teach higher-level science content to emerging colleagues while understanding important lessons about science communication both from the perspective of a scientist and a pedagogue. The collaborative co-teaching model allows the course instructors to "try out" different identities in ways that improve their respective roles in urban science teaching. The "trying out" eventually leads to greater awareness of learning and the integration of changes that make it more relevant to effective urban science teaching and learning.

The co-teaching of the courses provided the classroom teacher/instructors the opportunity to reinforce what they learned during their own coursework and to continue to deepen their content knowledge and place-based practice. They interacted with other classroom teachers (both co-teachers and in-service teachers in the course) and faculty (co-teacher and guest lecturers), and through these interactions they position themselves as place-based classroom teachers through the teaching and learning and adding layers of knowledge about local places. A collaborating teacher discussed her experience on a field trip,

> Wayne (guest lecturer) had added a lot of historical and architectural information to the geology information we'd learned about when I had first taken the tour [in the course], so I learned that. I also used the review of how to look at rocks in building stones to enhance my understanding of common rocks. I'll have more background knowledge when I teach my students about common rocks in NYC.

Wayne, as a faculty member, continues to research and build his knowledge about local rocks and stones and shares this knowledge with teachers through co-teaching and leading field trips. This then becomes shared knowledge in the community of City-as-Lab educators as they share this knowledge with students in their classroom, allowing them to focus their science lens with which they view their city.

Developing a shared practice. The co-teaching model afforded the space for the geoscience faculty to develop shared resources—activities, language, practices—for facilitating the geoscience courses for teachers and under-graduates. A key issue that arose was developing grade-appropriate student projects. For the faculty, the experience helped him to realize the breadth of content knowledge that teachers need depending on the grade level. With the help of his collaborating teachers, he was able to develop grade-appropriate projects and create appropriate assessment tools. Prior to working with the teachers, he felt that some of the projects completed by the elementary-level teachers showed evidence of only a minimal grasp of the subject; however, being able to see it thought the lens of an elementary classroom teacher, he was able to view and assess the projects as relevant to their grade level. His collaborating teacher concurred, "We discussed what were "grade-appropriate" products for the Building Stones of NYC [project]. Students created lesson plans with products such as picture books, interactive games and trip slip/worksheets for their own students to use in school." In this case, she positioned herself as the pedagogical specialist in place-based science learning and described her role as helping the professor to create a rubric to use "for the very different submissions based on [my] experience with elementary-age children's needs." As this was a key project in the course, the other collaborating teacher cited:

> We discussed the kind of product that would be appropriate for the GEO 613 students who are education majors studying Earth science/geology to do. We decided that elementary teachers should create a picture book teaching the building stones of lower NYC to their students. We would write the assignment with a rubric and [collaborating teacher] and I would help [geoscience faculty] grade it according to education standards as well as science standards. Middle/high school teachers would create a power point lesson they could use in their classrooms.

The work around developing the projects and rubrics represent the shared meanings and tools that the City-as-Lab holds around appropriate artifacts and assessments to demonstrate understanding in the students who take the course. It is through interactions that these shared practices and meanings emerge and these shared meanings and practices shape teaching identities.

The collaborative environments are also important in fostering place identity. In revisiting the course themes (see Table 15.1), both classroom teachers and faculty collaborators design syllabi and teaching activities that connect the themes to places. Through the collaborative relationships, the course instructors develop place identities connected to using specific sites as resources for their courses as they return to these sites year after year to enact particular activities. Whereas place identity often implies a connection to a specific physical and geological site, for educators, it is about developing a place-based practice and an identity with place-based practice that allows for the recreation of place-based experiences in different similar sites. For instance, whereas the resources in a major art museum are central to the Global Arts and Artifacts course, teachers could use a museum with similar artifacts that is more accessible to the school or bring place into the classroom through examining global artifacts and art in the classroom. This is also an opportunity to leverage place, identity, and living in a diverse urban center—students could contribute and/or recreate artifacts that are relevant to the place identities that are connected to their notions of "home" and ethnicity.

Continuity of the City-as-Lab Community

We continue to identify as place-based educators and expand our City-as-Lab model through the development of new courses, grant opportunities, faculty research, and our participation in a community of science educators ranging from pre-K through college faculty. We maintain our collaborations with informal science institutions and maintain our view of the community as an extended learning context. Although we focus on local resources, we afford a teaching identity that emphasizes place and is transferable to other contexts, we often receive emails from teachers who move to a different city and seek out similar resources to maintain their teaching and place identities in a new place.

Two recent experiences demonstrate the impact of our programs on the local earth science education pipeline. A chance conversation with an incoming student in our City-as-Lab graduate program, a recent graduate of our undergraduate earth science teaching program, revealed that her high school earth science teacher and mentor was one of the first graduates of our City-as-Lab master's program. This student represents an aspirational outcome of City-as-Lab—a "home grown" teacher of color, with an immigrant background, who will most likely stay in the urban classroom. At a high school research symposium, we were introduced to a Latina Brooklyn College environmental studies major who had just completed her freshman year. The teacher, a City-as-Lab graduate, organized the symposium and invited this former student to return as a guest scientist. Looking around her the student noted, "This is what all science classrooms should be like! (Students actively engaged in doing science.)" Both are examples of program graduates inspiring students to become earth scientists and to study at Brooklyn.

Because urban science teaching is often, unfortunately, viewed from a deficit perspective, it strengthens our resolve to shift the perception to being one of promise in access to an almost endless range of science-rich places only constrained by time, curricula, and school policy. Some teachers in our program begin to view place-based learning as an equity issue in that exposure to the city through a science-rich lens expands opportunities for student learning. For them, field trips and place-based learning are essential to providing their students rich, meaningful, and diverse learning experiences and opportunities for them to expand their realm of future possibilities. Through place-based teacher learning experiences, teachers realize firsthand the power of learning in place and seek to recreate similar experiences for their students.

The City-as-Lab learning community provides the context for educators, both classroom teachers and faculty, to develop science teaching identities connected to place-based practices. Through these place-based practices, they develop place identities that are connected to viewing urban science teaching and learning as full of potential and opportunity to engage in deep, meaningful, and relevant science learning. Also, because as we learn about places we also teach places about ourselves, educators are empowered to recreate places as science rich for the enactment of their own place and teaching identities and for context-relevant science learning opportunities for their students.

REFERENCES

Avraamidou, L. (2014). Developing a reform-minded science teaching identity: The role of informal science environments. *Journal of Science Teacher Education, 25,* 823–843.

Adams, J., & Gupta, P. (2013). "I learn more here than I do in school. Honestly, I wouldn't lie about that": Creating a space for agency and identity around science. *International Journal of Critical Pedagogy, 4,* 87–104.

Gross, M., & Hochberg, N. (2014). Characteristics of place identity as part of professional identity development among pre-service teachers. *Cultural Studies in Science Education.* doi: 10.1007/s11422–014–9646–4

Kudryavtsev, A., Stedman, R., & Krasny, M. (2012). Sense of place in environmental education. *Environmental Education Research, 18,* 229–250.

Powell, W. (2005). *Teaching urban students.* Retrieved from: http://serc.carleton.edu/NAGTWorkshops/teaching_methods/urban/index.html

Tobin, K., Roth, W.-M., & Zimmermann, A. (2001). Learning to teach science in urban schools. *Journal of Research in Science Teaching, 38,* 941–964.

van Eijck, M., & Roth, W.-M. (2010). Towards a chronotopic theory of "place" in place-based education. *Cultural Studies of Science Education, 5,* 869–898.

16 The Promise of Collaboration
Classroom Teachers and the Use of Informal Science Education Resources

James Kisiel

> My father was a Methodist minister, but he was going to be a scientist. He knew all the astronomy science, every star. . . . He could tell you about every plant, native to California. We were always with nature. . . . For me, [I was] growing up with all of this around me, and then when I had my children, I kept taking them to science museums and taking them to the beach and the mountains, just to go see everything. My daughter is now working as a soil scientist.
>
> (K, elementary teacher)

> It is really important to me, that these kids have exposure to a lot of interesting things in their lives. Unfortunately in my school, my principal is pretty much anti-anything outside of the book. She is very much "Learn from the book. We don't need science. If these kids can't read, why should they learn science?"
>
> (N, elementary teacher)

Learning is complex. Where we go, with whom we interact, what we see and do—these are just a few of countless factors that can influence how we make sense of the world. As we see throughout this volume, learning is not bound to school; yet for many, at first thought, learning is synonymous with schooling. Such demarcations, even if unintended, truly limit our ability to support learning in ways that ultimately contribute to an informed citizenry.

Within the realm of science education, the idea of leveraging resources from the out-of-school or informal learning sector (e.g., museums, aquariums, nature centers, or summer camps) continues to gain support from such organizations as the National Science Teachers Association and the Institute for Museum and Library Services. The National Research Council (2009) suggests that "learning experiences across informal environments may positively influence children's science learning in school, their attitudes toward science, and the likelihood that they will consider science-related occupations or engage in life-long science learning through hobbies and other everyday pursuits" (p. 304).

Despite a growing number of similar endorsements, the pairing of the formal and informal learning resources has proven to be challenging (e.g., Kisiel, 2014). Formal collaborations between institutions, while potentially fruitful, frequently seem to be marriages of convenience, spurred by funds or top-down mandates. Such institutional interactions require time for participants to develop new practices that allow stakeholders from both institutions (e.g., school and museum) to fully utilize a new suite of educational resources. Unfortunately, such time and support may not be adequate, resulting in a kind of "one more thing on my plate" perspective from the participating educators.

In many cases, however, the individual teacher leads the interaction between informal science education institutions (ISEI) and schools (Bevan & Semper, 2006). She or he must negotiate between ISEI and school settings to discover and access such resources, effectively blend them into the curriculum, and ultimately create a successful learning experience. Numerous studies of school field trips suggest that teachers struggle with the use of informal settings in supporting classroom pedagogy, and may not recognize how best to utilize such resources. Despite increasing pressures of student learning accountability and decreasing financial resources, we know that there are *still* those teachers who in fact do utilize these community resources to support their science instruction. Who are these teachers who are able to work successfully at the boundaries of the formal and informal settings? What makes them more likely to conduct field trips, use science curricula, or attend professional development sessions offered by ISEIs? Understanding how the contexts of school and ISEI, as well as personal experiences and perspectives, influence these "avid users" may provide insights into how best to foster collaboration to improve STEM teaching and learning.

BOUNDARY PRACTICE AND IDENTITY

Lave and Wenger (1991) describe a community of practice as a set of relationships between individuals, their activities, and those in other external groups. These communities share purpose, tools, and particular practices that contribute to a unique culture. Such a perspective allows us to reframe the interactions between school and ISEI (such as a field trip) as what Wenger (1998) refers to as a boundary practice—an activity that involves interactions of two or more communities of practice. He reminds us that a community of practice is as much defined by commonalities within (practices, objectives, tools) as it is by those things that distinguish it from other communities. The boundary defines what the community of practice is and what it is not. Yet as part of our practice, we are constantly interacting with other communities of practice. Not only that, but we are members of multiple communities. Part of who we are (i.e., our identity) is based on how we resolve our multi-membership in different communities of practice, as well as how we interact with those communities for which we are not a member.

A *boundary practice* is one way that two communities interact through practice. It would seem that most typical school/ISEI interactions are based on a common activity (or practice): the field trip, the teacher workshop, the outreach program, etc. This boundary practice is unique (compared with the native practices of teacher and informal educator) and is established as an ongoing activity—one that retains characteristics of both communities of practice. The course (and perhaps success) of these boundary practices then relies, in part, on the abilities of brokers, members from one or both communities, to understand and interact with those from the other communities. In an earlier study of an aquarium-elementary school collaboration, a few teachers and an aquarium educator) played key roles in helping the two communities (elementary teachers and aquarium educators) better communicate common objectives and develop tools that could be used by both (Kisiel, 2010). That interaction differed from the typical school-ISEI interaction in that interactions were explicitly developed for those two specific communities and were sustained over the course of several years—such sustained interaction led to noticeable changes *within* each community practice. In the case of the field trip or outreach program, we see a particular activity that remains outside the norm of practice. In these cases of infrequent interaction, the importance of the broker becomes even more important, as the interactions are typically finite and gain a kind of special event status beyond typical practice—especially for schoolteachers.

Wenger (1998) also speaks of the importance of identity in understanding those who participate in communities of practice; the way an individual defines herself or himself is influenced by community membership (and nonmembership) as well as the nexus of multi-membership or the ways we reconcile different memberships into a singular identity. Identity is in part formed through participation, "not just through reified markers of membership but more fundamentally through the forms of competence that [membership] entails" (p. 152). Identity tends to be seen as a certain kind of person, who is a function of context external to the person. Identity is recognized through a combination of individual's speech, personal interactions, beliefs and values, use of objects or tools, and so on.

Although the definition of identity varies across disciplines, it is a combination of external and internal contexts that shape how someone sees themselves. The individual might be seen as an active agent, driven by all the experiences, ideas, obstacles, and perspectives accumulated via participation over time. Whether this entire collection of characteristics might be labeled as the individual's *identity* may be less important than determining whether that combination of characteristics might be indicative of future agency. Are some of these components more important than others? Or, is the whole so integrated that one aspect (an experience, an attitude) cannot be separated from the rest?

Examining both the internal (personal) and external contexts of avid users may provide us with a profile of a kind of teacher who can find reconciliation between two different communities of practice—that of the school and that of the museum or other informal institution—so much so that using ISEI resources is not challenging or counter to their normal practice as teacher. From this perspective, it would seem that avid users would serve Wenger's role as brokers for other boundary encounters. If this is the case, then identifying these key characteristics should provide important suggestions for improving overlap between two sometimes-disparate communities of practice.

THE INVESTIGATION

To understand these avid users better, an exploratory investigation was conducted to identify key factors, within both personal and institutional contexts, that might influence avid users (those teachers who report higher levels of ISEI use) and their efforts to support science learning through the use of informal science institutions. Before describing this process, is important to understand that frequency of use is not necessarily an indicator of the *quality* of the learning experience provided by the teacher, and teachers who engage in such activities only occasionally may indeed conduct that effective use of ISEI resources. As such, one limitation of this investigation is that the effective user (defined any of a number of ways) is not identified explicitly through this analysis. Yet despite this limitation, it would seem that the avid-user teachers who were the subject of this investigation found value in using these institutions, and given the complexity of coordinating such learning experiences, managed to circumvent some of the obstacles that might prohibit other teachers' participation. Their perspectives should provide valuable insights regarding how best to improve teacher access to and utilization of such experiences.

Three informal science institutions in Southern California served as focal points for the investigation (an aquarium, a science center, and a botanical garden). Teachers who had engaged in some way with each institution—through a field trip, professional development program, outreach program, etc.—were contacted via institutional email or traditional mailing lists. Teachers were asked to complete an online survey regarding their prior experiences, attitudes toward ISEIs and science teaching, perceptions of challenges regarding use of ISEI resources, and some basic demographic information regarding their school and training. Such items were derived from other studies and findings reported in the field trip and informal science learning literature. In addition, confidence in science teaching was also examined through implementation of the Science Teaching Efficacy and Beliefs Instrument (Riggs & Enochs, 1990). Responses from 235 teachers were obtained across the three sites. Teacher focus groups and interviews were also conducted with selected respondents in an effort to clarify trends observed in the quantitative data.

Results from the teacher survey led to a collection of variables at different measurement levels that reflected teacher attitudes toward science, school, and informal learning institutions, as well as other characteristics and prior experiences. Initial analyses (including Spearman correlation and chi-square) of these and other variables from the survey were used to simplify this complex data set and remove factors that were highly correlated with each other, resulting in a set of 31 predictor variables that were used for further analysis (Table 16.1).

Table 16.1   Predictor variables used for avid-user analysis.

School characteristics (dichotomous variables)
Designation Title I
Designation No Child Left Behind "needs improvement"
Charter or private school

Perceptions of field trip success (dichotomous variables, based on open-ended question)	
Student outcomes	Success defined in terms of student learning and other student-related outcomes (if any of the following categories were identified, then this was also coded).
Positive experience for students	Mention of student enjoyment, excitement, and 'having fun'.
Evidence of student learning	Explicit mention of learning, understanding, or new knowledge.
Linked to curriculum	Mention of links to or alignment with classroom learning or content standards.
Fostered student interest	Mention of tapping student interest and ongoing discussion during and after the experience; connecting to students' lives or interests.
Students exposed to new experiences	Describe experiences as new to students or opportunities they might not otherwise have; students exposed to broader perspectives not found in school.
Operational outcomes	Success defined in terms of logistics, preparation, execution (if any of the following categories were identified, then this was coded).
Engaged students	Reference to students being engaged, on task, or attentive; references to students' good behavior or lack of discipline problems; no incident.
Good ISEI programming	Reference to quality of programs (tours, classes, etc.), staff, or volunteers; mention of any museum-led activities.
Teachers/students well prepared (good planning beforehand, etc.)	Mention that students (or teachers) are prepared or organized.

(Continued)

Table 16.1   (Continued)

ISEI support for teachers (dichotomous variable, based on open-ended question)	
Provide funding	Providing scholarships or funding for admission or transportation.
Provide training	Professional development or other instructional assistance.
Provide outreach to schools	Having the institute come to the school (often in lieu of a field trip).
Provide additional support for field trips	Providing materials (maps, pre-/post-visit lessons, chaperone guides, etc.).
Provide instructional materials	Curriculum guides or teaching materials, NOT necessarily related to field trips.

Teacher characteristics	
Training-high	Teacher indicated a "fair amount" of training related to ISEIs as part of education or certification; those who responded "a little bit," "not really," or "don't remember" were considered *training-low* (dichotomous).
Teaching experience (in years)	Open-ended response (continuous).
STEBI score	Based on validated instrument designed to measure science teaching self-efficacy, primarily for elementary teachers (continuous).
PSTE score	Subscore from STEBI instrument: Personal Science Teaching Efficacy relates to the teacher's sense that their science teaching can impact the students they work with (continuous).
STOE score	Subscore from STEBI instrument: Student Outcome Expectancy relates to the teacher's sense that students are able to succeed in (i.e., learn) science class (continuous).
Frequency of field trips as an *elementary student*	Responses on a five-point, self-report scale (ranged from more than once per year to never [ordinal]).
Frequency of field trips as a *secondary student*	Responses on a five-point, self-report scale (ranged from more than once per year to never [ordinal]).
Variety of sites visited as student	Number of sites checked from a list of 20 different sites/settings (museums, aquariums, parks, farms, etc. [continuous]).
Frequency of field trips *with family as child*	Responses on a five-point, self-report scale (ranged from more than once per year to never [continuous]).

Teacher characteristics	
Variety of sites visited *with family*	Number of sites checked from a list of 20 different sites/settings (museums, aquariums, parks, farms, etc. [continuous]).
Role as trip organizer	Reported organizing and leading trips for their own class OR for theirs and classes of other teachers (based on multiple choice). Other responses (organize as team, another teacher organizes) were grouped as 'non-organizers' (dichotomous).

Challenges (ordinal variables, five-point scale ranging from strongly agree to strongly disagree)	
Time	Time prohibits ability to organize a field trip.
Linking to curriculum	Linking field trips to standards/curriculum is difficult.
Unsupportive administration†	School admin is generally *not* supportive.
Funding not available†	School is *not* able to provide funds for field trips.
Student behavior	Difficult to control student behavior.
Chaperones	Difficult to find chaperones.
Logistics	Logistical details.

† These items were reversed in the survey, described as positive supports.

Initial analysis of elementary and secondary teacher responses, via simple nonparametric correlations, suggested that these two groups had different perspectives regarding their use of ISEIs. This is not particularly surprising given the different instructional contexts each group faces (e.g., self-contained classrooms vs. multiple subject-specific classes). Because such differences in practice are likely to impact teacher perspectives, it seemed prudent to think of the sample as two groups—elementary and secondary teachers. Each group was examined separately, leading to separate (although possibly overlapping) profiles based on the selected variables.

Making sense of this rather messy collection of continuous, ordinal, and categorical or dichotomous variables required the use of a General Linear Modeling (GLM) approach. GLM is a type of regression analysis that examines the influence of multiple variables on a particular numeric dependent variable (in this case, frequency of use.) The approach is more flexible than typical regression analysis in that it allows for all types of variables (e.g., continuous and categorical).[1]

WHO ARE THE AVID USERS?

At the start of this investigation, the avid-user teacher was viewed at that teacher who looked beyond the classroom to see all of the different science teaching resources throughout the community—zoo field trips, natural history museum outreach programs, aquarium speaker series, etc. Yet this assumption of broad use turned out to be faulty. In fact, analysis revealed a low correlation between a teacher's frequency of field trip activities and their use of all other ISEI resources mentioned in the survey (outreach programs, teacher PD workshops, web-based resources, science lectures, and materials/kits). Furthermore, the frequency of use of other ISEI resources was generally highly (and significantly) correlated. This would seem to suggest that there are essentially two kinds of avid users—those who frequently use field trips to engage their students and those who frequently utilize other community resources. Analysis, then, was refined to look at avid users in terms of two different dependent variables, rather than a conglomerate of field trip use with all other possible uses. This led to four different analyses and the development of four different teacher profiles that might help us understand what factors and self-perceptions are likely to contribute to a teacher's engagement with ISEIs and out-of-classroom resources (Table 16.2).

Four separate GLM models were developed using the variables described earlier. As expected, each of the models developed was distinct, although several similarities did emerge across all four teacher profiles. GLM results are provided in the appendix at the end of the chapter.

THE ROLE OF INSTITUTION AND EXPERIENCE

Institutional Contexts

The analysis of this extensive data set reveals four distinct profiles for teachers who are frequently able to look beyond the classroom walls to resources and opportunities within the community. The complexity of this analysis reminds us of the complexity of contexts within which teachers practice. Despite this complexity, the analysis does point to several characteristics of the school setting that seem to affect how avid users participate in these formal-informal boundary encounters. Data suggest that avid users are

Table 16.2   Four profiles of avid users.

	Elementary teacher	Secondary teacher
Avid *field trip* users	Profile 1	Profile 2
Avid *other-resource* users	Profile 3	Profile 4

generally more likely to work in private or charter schools, and although accountability pressures and curriculum requirements were described by all teachers in both surveys and interviews, the analysis suggests that avid users were less likely to see these factors (especially the extra time commitment needed to conduct a field trip) as deterrents.

In addition, it is important to note what did not seem to be indicative of avid users. The availability of school funding for field trips, a "passing" NCLB status, or an affluent student population (as indicated by lack of Title I designation) were not found to be institutional prerequisites for avid users in comparison to their peers who used such resources less frequently. Whereas these might simply be seen as negative correlates to the private/charter school affiliation, their lack of importance does point to the fact that these commonly identified challenges do not inhibit these interactions for avid users. What needs clarification, however, is whether the absence of challenges is due to relaxed pressure from the institution (charter/private school) or an internal drive or agency that the avid user brings to bridge the formal-informal boundary.

Personal Contexts

Several internal factors or personal characteristics were explicitly found to be indicators of avid users. Field trip avid users were more likely to *define field trip success in terms of student outcomes* (students learned, students had a positive experience, students were exposed to new things) rather than operational outcomes (buses were on time, programs were good, volunteers were effective, etc.), suggesting a difference in their objectives for the experience or possibly differences in concern related to logistics. It is worth noting that for some profiles it was not that the teachers' defined success in terms of student outcomes, but rather that they did not define success in terms of operational outcomes.

Prior experiences related to museums seemed to play a role, especially for the secondary avid-user teachers. They reported a greater variety of family museum experiences, fewer field trip experiences as high school students, and prior training in the use of ISEIs. The importance of a lack of field trip experiences is worth a closer look, as it suggests that their use of ISEI resources is a reaction to their limited experiences as a student. One of the avid users explained her rationale during a follow-up interview:

> I think having gone through a school system where I had very few field trips and those hands on experiences [was difficult]. I loved learning and I was always interested, I loved reading so for me, I wanted to know more. I wasn't given the resources. As I got older I was able to do things for myself. Like, "Wow, there is an observatory over there!" It opened a whole new world for me. I guess that is what I sort of go back to in my teaching. I want to inspire.

Science teaching self-efficacy, as measured by the STEBI instrument, was another factor identified as being potentially influential in the decision to use ISEI-based experiences to support learning. For secondary science teachers, the STEBI subscore, Personal Science Teaching Efficacy, or the teacher's perspective that they can impact student learning, was identified as an important factor related to frequent use of non-field trip resources. Such a link might be expected to be a strong indicator across all profiles of avid users, as it would suggest that the teachers' confidence in impacting students learning drives their efforts to find resources beyond the classroom. The fact that this is not seen for either of the elementary profiles might be indicative of the multidisciplinary nature of the multiple-subject classroom. Ultimately, the teachers' efficacy in another subject area (e.g., history, language arts, etc.) might drive their desire for external resources (whether they are science-related or some other discipline). It is also worth noting that the other STEBI subscore, Science Teaching Outcome Expectancy, a measure of the teacher's perspective that students will learn or benefit from science teaching in general, was found to have a significant but negative coefficient, suggesting that lower STOE scores were a stronger indicator than higher STOE scores. This apparent contradiction might be explained in terms of a compensation effort, similar to how a lack of field trip experience encouraged teachers to do more. In the case of the outcome expectancy scores, it may be that concerns over the impact of classroom science instruction in general may spur teachers to look beyond the classroom for ways to improve relevance or tap into other student interests.

As with the analysis of institutional context, several personal characteristics surprisingly did not distinguish avid users and their counterparts. Most notable was the fact that classroom experience (or number of years as a teacher) did not seem to have any relationship with whether or not a teacher was likely to be an avid user. Whereas this may seem contradictory, we must remember that the use of a community resource like a zoo or science center is generally not a standard part of teaching practice. Whereas teachers may develop expertise in classroom management, assessment, or student engagement over time, looking beyond the school for ways to supplement or enrich instruction may simply be outside the norms of practice. Certainly, the practices that developed as a result of the assessment-laden No Child Left Behind legislation reset the norms of teaching practice in ways that in some ways decontextualized school learning, essentially separating it from the community. Thus it would seem that other factors, such as those mentioned earlier, are simply more influential than teaching experience.

IDENTITY AND THE AVID USER

Emerging Characteristics

The GLM analysis provides a statistical representation of factors that appear to be important in describing who these avid users are. We must realize the limitations of such analyses in their dependence on clear definitions and measures of variables as well as the importance of sample size in determining the level of influence for each of the different variables. For all profiles, we must look at the defining characteristics as a first approximation—a larger sample size would likely strengthen our confidence in variables identified and might even lead to additional variables that were initially deemed insignificant.

It would seem, though, that this exploration into the personal and institutional contexts that potentially influence teacher-ISEI boundary interactions provides important insights for fostering such interactions. Although each avid-user profile is distinct, there are several categories of variables that appear to characterize teachers who are likely to use informal science resources more frequently than their peers.

First, we see the importance of *key experiences*. In all but one teacher profile, museum-related antecedent events, whether prior field trips as a student, family trips, or even interactions (training) as part of teacher preparation, were identified as an influential factor related to repeated use of informal science resources and programs. It seems reasonable that such firsthand experiences might help teachers develop a value for the ISEIs as learning institutions. The quote at the opening of the chapter exemplifies how one teacher's experiences as a child led to a lifelong appreciation of out-of-classroom learning experiences, even contributing to her child's decision to go into a science-related career.

A second characteristic suggested through the development of these teacher profiles is the importance of *instructional responsibility*. The idea that student-centered outcomes are indicators of field trip success in several of the models may suggest that for these teachers, ISEI use supports their goals as teachers. During the interviews and focus groups, teachers clarified why field trips and similar activities were important:

> [Because] I know I gave everything I could to provide a good education for that student. If I suddenly sell short of that and I just allow her to learn [only] in a classroom setting, I am depriving them of that. I am making it a choice to go one step further in giving them the best education they can. For me, going on a field trip or bringing out my science kit . . . is allowing them to become even better than they were five minutes ago.

This perspective was common among avid users (i.e., "We do it because it's just good teaching.") and also helps us explain why obstacles such as costs or time were not important for these teachers.

Closely tied with the idea of instructional responsibility is the importance of *teacher agency*, as evidenced by many of the contributing factors across the different models. The fact that commonly cited obstacles such as funding, transportation, curriculum connections, and time are not prominently featured in these profiles points to teachers who are taking it upon themselves to make use of ISEI resources. This sense of agency allows them to circumvent or push through what many of their peers see as impassable.

We see in this investigation how many different factors can influence avid use of ISEI resources. Elements of teacher identity—agency, responsibility, and experience, for example—manifest themselves in the way that both personal perspectives and external opportunities (and limitations) affect their abilities to engage in these out-of-school resources. Whereas identity might be conceived as an individual trait, it is difficult to know whether the identity was shaped by these factors or whether the relevance of these different factors was determined by a well-defined identity. It is this interaction between internal and external that makes the concept of identity somewhat fluid and hard to define. Nevertheless, the data here does suggest that there is something about "those teachers" who are determined to navigate the space between informal and formal learning settings.

Avid Users and Boundary Agents

Using the community of practice lens, we now have a clearer idea of what it might take to work across several communities of practice and participate in a boundary activity. Wenger speaks of *brokers* as having membership in multiple communities of practice, bringing elements of one practice into another. Although avid users may not strictly have membership in both school and museum communities of practice, it would seem that they may better understand the different communities of practice of the museum, perhaps serving as boundary agents. Not only have these teachers managed to navigate boundaries within their own institutions but they also managed to interact successfully with some of the other communities of practice within museums (e.g., reservationists or educators). This does not suggest that these teachers are the only ones who *should* participate in such boundary activities, but rather they may be best equipped to identify the challenges in engaging in that activity, and as such, shepherd others, whether from school or museum, across that boundary.

At the heart of this investigation lies a clearer understanding of avid users and the relevance of key experiences, instructional responsibility, and agency in shaping these teachers' decisions of frequent participation in activities at the boundary between school and museum. Although the findings point to external influences that afford teacher engagement with these community science resources, they also point to the internal components of teacher identity as an important contribution to their use of ISEIs. The findings of this study have implications for both informal science education institutions as well as teacher educators. If we consider the notion that identity is not static and that it can evolve with time, experience, and the interaction of both internal and external contexts, then several possible strategies might be considered. The introduction of new tools, such as field trips, outreach programs, or museum-based professional development workshops, may help familiarize teachers with practices common to informal settings. Such tools and training may also foster development of identity components, such as "one who uses outreach programs" or even "elementary science teacher." As teachers begin to participate in new boundary practices, their sense of agency and sense of belonging further strengthen an identity as an ISEI-user.

If we look to avid users as boundary agents or boundary crossers, we must also keep in mind that their developed expertise may still be limited. The analysis described here points to the necessity of separating elementary and secondary teachers into separate groups, as the structure of their school-based practices differ. The challenges and affordances of a multiple-subject classroom make an elementary teacher's expertise for using ISEI resources much different than that of a high school biology teacher. But even more important is the discovery that a teacher's frequent use of field trips does not imply an awareness or use of the *other* resources these community sites have to offer. The observation that this group of teachers identified as avid users can be divided into avid field trip users and avid other-resource users reminds us that this intersection between formal and informal is a complex one. Not all teachers are the same. Not all teachers have similar goals, experiences, or rationales. And of course not all teachers see their classroom in the same way. Assumptions about teachers, learning opportunities, and collaboration need to be reexamined by all stakeholders—informal educators, school and museum administrators, parents, and teachers—if the desire to support science learning as a community-wide effort is truly important.

NOTE

1   A manual backward elimination approach was used, whereby all factors were initially loaded into the model. Examination of the first iteration led to the removal of variables with the highest p-values ($p \geq .900$); then the calculations were re-run. This stepwise removal of factors continued until $p \leq .100$. This process resulted in the final models.

REFERENCES

Bevan, B., & Semper, R. J. (2006). *Mapping informal science institutions onto the science education landscape*. Retrieved from: http://www.exploratorium.org/cils/research.html

Kisiel, J. (2010). Exploring a school–aquarium collaboration: An intersection of communities of practice. *Science Education, 94*, 95–121.

Kisiel, J. (2014). Clarifying the complexities of school-museum interactions: Perspectives from two communities. *Journal of Research in Science Teaching, 51*, 342–367.

Lave, J., & Wenger, E. (1991). *Situated learning: Legitimate peripheral participation*. Cambridge, UK: Cambridge University Press.

National Research Council. (2009). *Learning science in informal environments: People, places and pursuits.* Board of Science Education, Center for Education. Washington, DC: The National Academies Press.

Riggs, I. M., & Enochs, L. G. (1990). Toward the development of an elementary teacher's science teaching efficacy belief instrument. *Science Education, 74,* 625–637.

Wenger, E. (1998). *Communities of practice: Learning, meaning, and identity.* New York, NY: Cambridge University Press.

Appendix
General Linear Modeling Analysis Results

Table A.1 Contributing factors for avid field trip use, elementary teachers (Profile 1).

Variable	Coefficient (B)	p	Effect size	Observed power
Organizer	0.438	0.013	0.093	0.714
Success-student outcomes	0.846	0.016	0.087	0.679
Charter or private school	0.511	0.022	0.079	0.638
Time limitations NOT prohibitive	0.204	0.047	0.06	0.513
Success-preparation and planning	0.648	0.065	0.052	0.457

Note: R^2 (adjusted) = 0.274

Table A.2 Contributing factors for avid field trip use, secondary teachers (Profile 2).

Variable	Coefficient (B)	p	Effect size	Observed power
Charter or private school	1.223	0.000	0.47	0.996
Variety of sites visited with family	0.138	0.000	0.429	0.989
Frequency of field trips as secondary student	-0.325	0.004	0.277	0.86
Success-operational outcomes	-0.622	0.006	0.257	0.824
School NOT able to provide funds	0.282	0.006	0.257	0.822
NCLB "needs improvement"	0.657	0.007	0.245	0.798
Success-curriculum link	0.578	0.027	0.174	0.615
Science Teaching Outcome Expectancy (from STEBI)	-0.041	0.034	0.161	0.576
Success-fosters student interest	0.591	0.061	0.128	0.47

Note: R^2 (adjusted) = 0.552

Table A.3   Contributing factors for avid other-resource use, elementary teachers (Profile 3).

Variable	Coefficient (B)	p	Effect size	Observed power
ISEI assistance-funding	0.441	0.012	0.088	0.718
Variety of sites visited w/family	0.062	0.012	0.088	0.721
NOT difficult to find trips linked to curriculum	0.163	0.024	0.072	0.624
Time limitations NOT prohibitive	0.176	0.07	0.047	0.442
Success-operational outcomes	-0.282	0.099	0.039	0.379

Note: R^2 (adjusted) = 0.268

Table A.4   Contributing factors for avid other-resource use, secondary teachers (Profile 4).

Variable	Coefficient (B)	p	Effect size	Observed power
Received training re: ISEIs	1.166	0.001	0.341	0.956
Personal Science Teaching Efficacy (STEBI)	0.063	0.001	0.348	0.962
NCLB "needs improvement"	1.024	0.002	0.294	0.909
ISI assistance-training	0.805	0.003	0.276	0.883
ISI assistance-field trips	-0.57	0.005	0.247	0.833
Title I school	-0.658	0.017	0.187	0.688
Variety of sites visited w/family	-0.066	0.031	0.155	0.592
Charter-private school	0.482	0.057	0.123	0.482
School admin supportive	0.138	0.12	0.084	0.341

Note: R^2 (adjusted) = 0.498

ACKNOWLEDGMENTS

This work was made possible through funding from the Spencer Foundation and the cooperation of many classroom teachers and informal educators who generously contributed their time and perspectives.

17 Stories of Self and Informal Science

Tracing Preservice Elementary Teachers' Identity Work Across Informal Science Experiences

Lucy Avraamidou

Reform recommendations around the world call for a reconceptualization of K–12 science education and reorganization of science learning environments. Consequently, these recommendations set high challenges for elementary schoolteachers, as well as teacher preparation. Hence there emerges a need to examine how elementary teachers come to be, how they learn, and how they develop through time and across contexts. Essentially, there is a need to explore the pathways by which beginning teachers come to develop, how they view themselves as teachers of science, how these views relate (or not) to reform recommendations, and what experiences shape these views. How teachers view themselves as teachers has been framed in the literature within the construct of *identity* (Avraamidou, 2014b). In this chapter, I suggest that as teacher educators aiming at preparing high-quality teachers, we ought to examine how beginning elementary teachers construct identities for science teaching and what kind of experiences throughout their lives are critical in shaping their identities. In so doing, I pay special attention to the *informal* science experiences that teachers have in various contexts throughout their lives. To frame this study, I use the construct of *identity work* and aim at responding to the following questions:

- What is the nature of two purposefully selected, beginning elementary teachers' identity work for science teaching?
- What informal science experiences throughout the participants' lives have influenced the nature of their identity work for science teaching?

In what follows, I first provide a set of conceptual and empirical underpinnings on science teacher identity and informal science learning, which frame this study. I then provide extracts from the personal stories of two preservice elementary teachers and illustrate how specific, informal science experiences throughout their lives have shaped their identity work.

INFORMAL SCIENCE

In recent years, a number of researchers and institutions around the world have shown interest in informal learning, which is generally used to refer to the learning that happens outside the school classroom and which operates across a broad range of contexts and disciplines, reaching out to people of all ages, in places such as museums, family, community, and other every day settings. In this chapter, I use the term *informal science experiences* to refer to the experiences that teachers have had either within formal or informal science settings, and which have the following characteristics: individuals choose with which activities they want to engage; there is no intended curriculum; activities are nonevaluative and noncompetitive; and social interaction among groups is heterogeneous and involves peers, adults, scientists, and informal science educators (Avraamidou, 2015). In a review paper, I summarize findings of existing literature about teacher preparation and informal science institution collaborations, which provide evidence that experiences in the context of informal science environments have the potential to support teacher learning and development in light of reform recommendation (Avraamidou, 2014a). More specifically, the outcomes of this review show that informal learning environments:

- Provide motivating activity structures;
- Provide opportunities to practice scientific inquiry and reform-based strategies;
- Offer opportunities to develop understandings about the nature of science and the work of scientists;
- Provide safe and fun learning spaces that are rich in resources;
- Have the potential to support the development of an appreciation for the value of science to society; and
- Have the potential to support the development of positive attitudes and orientations toward science.

(p. 840)

It is such arguments that have led me to explore the impact of informal science experiences, whether in or outside the formal school and university realms, on the development of preservice elementary teachers' science teaching identity.

TEACHER LEARNING AS AN IDENTITY CONSTRUCTION

The theoretical underpinnings of the argument I make in this paper are in agreement with Lave and Wenger's (1991) view that "one way to think of learning is as the historical production, transformation, and change of persons" (pp. 51–52). I hence use the term learning to refer to the construction of identities. Beijaard, Meijer, and Verloop (2004) argue that identity can also be seen as an answer to the questions: *Who am I at this moment*, and *who do I want to become*? The processes of being and becoming provide the theoretical basis of the argument I make in this paper, and I make a theoretical assumption that it is within social settings (e.g., universities, museums, community settings, family settings) and through interactions that beginning elementary teachers learn and develop and essentially construct identities for science teaching. To frame teacher identity, I use Lave and Wenger's (1991) social theory, which views identity as part of a social practice, and not just as an individual and isolated project. Central in this theory is the process of becoming, or the construction of identities. I therefore view teacher learning and development as a process of identity construction through social participation and am interested in the communities of practice that beginning teachers form in both formal and informal learning settings. The conceptualization of beginning elementary teachers' participation in various informal learning subcontexts as learners and future teachers of science is built upon Wenger's (1998) definition of participation, which "refers not just to local events of engagement in certain activities with certain people, but to a more encompassing process of being active participants in the practices of social communities and constructing identities in relation to these communities" (p. 4). The nature and characteristics of such participation in the practices of the social communities formed within the various informal contexts of learning-to-teach science is the point of interest in this study.

TEACHER IDENTITY AS A LENS TO TEACHER PREPARATION

The construct of identity has been receiving progressively more attention in science education in the past few years, as evidenced in the increased number of articles published in premier science education journals. As Varelas (2012) describes, "the multiple identities that students and teachers bring with them and further construct and reconstruct in classrooms and out-of-school settings allow them to be, and be recognized as, particular types of people" (p. 2). This framing of identity becomes useful in our understanding of and efforts to characterize teacher learning and development, especially as teacher educators around the world aim to address reform recommendations. In their review of the literature on professional identity, Beijard et al. (2004) identify the following features as essential for the professional identity of teachers:

- Identity is an ongoing process of interpretation and re–interpretation of experiences;
- Identity implies both person and context;
- Identity consists of sub-identities; and
- Agency is central, which refers to the need of teachers being active in the process of professional development.

In accordance with this conceptualization, "identity" is used in this paper to refer to the ways in which a teacher represents herself through her views, orientations, attitudes, emotions, understandings, and knowledge and beliefs about science teaching (Avraamidou, 2014a). Teacher identity is conceptualized as a sociocultural and tentatively shaped construct, constantly under development and always subject to change, given that a teacher lives in a social context where she interacts with others and where the influences of experiences are infinite. Given its cultural, social, and context-specific nature, identity becomes a valuable lens for studying teacher learning and development, especially within informal science settings. In this chapter, I examine the ways in which engagement in informal science programs and activities impacted two preservice elementary teachers' identity work for science teaching.

TEACHER IDENTITY AND FIGURED WORLDS

Researchers in the field of social anthropology have discussed identities as being formed through activity and participation within various cultural contexts of figured worlds (Holland, Lachicotte, Skinner, & Cain, 1998). These figured worlds feature as the central unit of analysis in this study, as they provide the "contexts of meaning for actions, cultural productions, performances, disputes, for the understandings that people come to make of themselves, and for the capabilities that people develop to direct their own behavior in these worlds" (p. 60). In line with studies about students' science identity work (e.g., Carlone, Scott, & Lowder, 2014), I view identity work as consisting of various figured worlds: (a) a figured world of family: experiences with science within the family environment; (b) a figured world of childhood: cultural and social influences related to science; (c) a figured world of schooling: events and experiences with science teaching and learning; (d) outside of school figured worlds: experiences with science outside of school; (e) a figured world of university: events and experiences with science teaching and learning at university; and (f) a figured world of science: understandings of and positioning (e.g., interest, appreciation, passion) in science. In conceptualizing identity work in this way, I aim to uncover critical events that the participants have experienced through their participation in various figured worlds, elucidate the meaning that they have assigned to those events, and explore the ways in which those events cause a shift to their science identities.

RESEARCH APPROACH AND PARTICIPANTS

I have used a longitudinal case study approach in order to explore the nature of two beginning elementary teachers' identity work for science teaching and the kinds of experiences that have influenced their identity work. Even though I have collected the data for this study over a period of three years, I apply biographical and narrative inquiry methods to collect data about the participants' early years of life. In this study, I attempt to collect various stories about the participants' experiences from their early years of life to their university years, by engaging them in various storytelling activities (e.g., biographical assignments, drawings, interviews, etc.).

A four-year-long teacher preparation program at a private university in a southern European country has served as the main context of this study. Two preservice elementary teachers (Maria and Alice—both pseudonyms) were selected to serve as focal participants in this study. Both were 18 years old when they entered the teacher preparation program, and they graduated from public high schools. Maria had specialized in science and scored high grades in her science courses, whereas Alice specialized in languages and got low grades in her science courses. For the purpose of this study, several kinds of data were collected for Maria and Alice: a biographical statement in relation to science, two drawings, a self-portrait, eleven journal entries, four lesson plans, a personal science teaching philosophy, and four one-hour-long interviews. The data were collected over a period of three years during which the participants were enrolled in the teacher preparation program. In organizing and analyzing the data, I developed portraits of each participant, which include biographical information and descriptions of their past stories as learners of science, as well as future stories as teachers of science, placing a special emphasis on the informal science experiences.

IDENTITY-WORK TRAJECTORIES OF MARIA AND ALICE

Maria

Maria can be characterized as an enthusiastic science learner, as evident through the analysis of her data set. She grew up in a small, rural village and remembers her childhood years being full of play. She loves teaching and looks forward to getting a job as a teacher and to having her own classroom. Science is one of her favorite courses because it has to do with everyday life, as it provides explanations about natural phenomena. I will share her personal narrative as constructed through various extracts from interviews and other biographical assignments related to her science identity work through time and across contexts, with an emphasis on informal science experiences. Her narrative illustrates her views about science teaching and learning, and the impact various experiences within the figured worlds of schooling, university, and family contexts had on her identity work.

> Central to my personal philosophy about science and learning is the use of informal science environments, such as museums, zoos, gardens, but also community settings or even the schoolyard. I believe that such approaches make science attractive and fun, but most importantly illuminate the value of science to society. They place science in specific contexts that are relevant to our everyday lives, such as, why it is important to recycle.

It is evident in this extract that central to Maria's personal philosophy for science teaching is the use of informal science environments and approaches. This view can be traced back to her experiences with science in various contexts. The following is an extract from Maria's interview where she recalls experiences in nature with her family:

> As a family we used to spend a lot of time in nature. I remember our Sundays spending long hours hiking in the mountains, and my father naming all the trees and then asking us to name them. We would play this game with my two brothers and always compete on who would remember the names of the trees.

When asked to share her memories from elementary school, Maria stated that she could only remember the science lessons held outside the classroom:

> My favorite memories from elementary school were the science lessons we had outside classroom—I remember one spring day when flowers were blooming and we walked to a path near the school to collect various types of leaves and flowers and brought them back to classroom to study them. I can't recall what we were actually doing in classroom but I remember how excited we were to be outside!

Elaborating on her memories from elementary school, Maria described teacher-centered approaches and negative experiences as a science learner:

> My memories of science at elementary and high school are negative. I found science very difficult to understand, and all my teachers were very strict and kind of weird. They used teacher-centered approaches, and we would always have to memorize information and do a lot of writing. It just didn't make sense to me; I couldn't understand why I had to know all those formulas. I did not see the value of science to my life.

Unlike her memories from elementary school as a science learner, her memories at university were positive. Similar to her personal philosophy, Maria stated that her favorite lessons were associated with informal approaches to science teaching:

> My favorite lessons during university were the ones I had in my methods courses, specifically the ones that were associated with informal science: the science fair, the water quality study, the visit to a natural history museum, and the waste management study that we did in the university's yard.

Elaborating on a couple of informal science experiences, Maria stated:

> I liked the museum and the botanical garden. I couldn't stop taking pictures of the animals, and I listened with great interest to the officer

who talked about the animals. Moreover, I enjoyed the activity at the botanical garden because I learned a lot about plants . . . I really enjoyed the herbal tea but I never knew which was what . . . now I do!

Talking about science fairs, Maria described how these could support students' development of positive attitudes toward science:

> I will definitely organize a science fair as a teacher because I believe it will help students develop positive attitudes towards science in the same way that it helped me. . . . It makes science learning fun, enjoyable and memorable because students are actively engaged in activities.

When sharing her beliefs about the role of informal science environments in science teaching, Maria made a comparison of such settings and the formal school context:

> I believe that informal science settings are very exciting and motivating for students mainly because they are social places, like an ordinary place. Classrooms are not natural settings; they are too organized, too structured. Unlike in the classroom, students were free to walk around and talk to each other and interact with different people!

As evident in her narrative, Maria has been an enthusiastic preservice elementary teacher who enjoyed science. As a young girl growing up in a small rural village, Maria's identity work as a learner of science was situated in informal settings with positive influences from her family figured world. She considered herself a successful learner of science and found science interesting and exciting, which provides evidence of how she positioned herself toward the figured world of science. As illustrated in her narrative, she had negative science learning experiences through her schooling years, except when science lessons were held outside the classroom. This illustrates the impact of the figured world of schooling on her identity work. In Maria's memories of science as a young learner, she pointed out that critical experiences during her teacher preparation were those of informal science: a science fair, a collaboration with a scientist to conduct a study about snakes, and an outdoors environmental study. Informal science approaches feature centrally in her personal philosophy about science teaching and learning, as she views herself as a future science teacher who will implement such approaches to her classroom practices.

Alice

It became apparent through the analysis of the data that Alice was a low-motive science learner in her former years, and her identity work was shaped by a series of negative science learning experiences. Alice grew up in the city without many experiences with science. She articulated no memories of science experiences within her family environment. As she said in the first interview, she was never really into science. I will offer several quotes from her interview, as well as relevant assignments that form her narrative in relation to her identity work for science teaching, placing an emphasis on experiences that took place within informal science settings. In her personal science teaching philosophy, Alice stated that informal science approaches featured centrally because of her own experiences with informal science as a learner at university:

> Informal science approaches to science teaching features centrally in my personal philosophy for science teaching. I experienced this approach as part of the methods course through readings and multiple informal science experiences, and I think it is really one of the best approaches for teaching science at the elementary school. The three informal science experiences I had at university (science fair, water quality study, interaction with a scientist) were the most fruitful ones. I learned so much because I was engaged, I found the concepts meaningful, I realized the value of science to society and I also learned about the work of scientists. In a way, these approaches helped me to understand how science is connected to our lives, and to reconstruct some stereotypical ideas I had about science and the work of scientists.

Sharing her memories from her elementary school years, Alice stated that she found science boring and uninteresting:

> Science was actually my least favorite subject from elementary through high school. It was boring and difficult to understand, I had absolutely no interest in it. Our teachers used teacher-centered approaches, and they all kind of had this "mad-scientist" look and personality. . . . weird and antisocial. I found science really boring in elementary school; it involved too much writing, actually copying what the teacher was writing on the blackboard. I remember always wondering, "Why do I have to know this?"

Discussing informal science approaches to science teaching, Alice pointed to the fact that informal science approaches provide opportunities to study science in context:

> I had never heard of informal learning and informal science approaches until I took the methods course. I believe it's one of the most valuable approaches a teacher could adopt, especially in science because you get to study things in context. For example, studying ecosystems or pollution in real contexts.

Explaining the value of informal science approaches to science teaching, Alice stated:

> Informal science approaches are so engaging because they fall outside the routine, which excites students. Moreover, they offer opportunities to student centered approaches, to students to explore their own questions, for example, when visiting a science center, they could choose to study what they want to, and spend more time at whichever exhibit they choose to.

Elaborating on the importance of adopting informal science approaches, Alice discussed how such approaches could support students in reconstructing stereotypical images about scientists:

> Prior to this class I was under the impression that scientists were old, strict men wearing lab coats and glasses. But I was wrong! This scientist was so young and cool, and had a great sense of humor. I loved listening to him! I will definitely invite a scientist to my class in the future to teach about snakes or any other topic. I believe that this is a very exciting and motivating approach for children because they rarely get to meet real scientists or to understand what science is really about and how scientists actually work.

Discussing her experience from implementing an informal science activity in the context of field experience, Alice discussed how excited the students were but also pointed to the various issues she had to face as a teacher:

> I tried an informal science approach during my field experience when I was teaching about animals, and we took kids to a nearby farm. First, I had to convince the teacher that this was going to be worthwhile, and then we had to get an official permission from the Ministry for the visit. This was a bit discouraging because it took a lot of time and effort. I didn't think I would have to actually convince them of how useful this would be to the children. The children were so excited and they kept asking about the day trip days in advance. We had engaged in some preparatory activities in the class, where they did some background research on specific animals so when we got there they already knew what they wanted to do. They worked in small groups based on their choice of what animals they would study, they made observations about their behavior, they collected data about their anatomy, and then interviewed the staff in order to collect more information about the animals. I had never seen them so excited, motivated and engaged in the activities!

In discussing how a series of informal science experiences in the context of the science methods course impacted her views about science teaching, Alice stated:

> I liked the museum and the botanical garden. I couldn't stop taking pictures of the animals, and I listened with great interest to the officer who talked about the animals. Moreover, I enjoyed the activity at the botanical garden because I learned a lot about plants. . . . I really enjoyed the herbal tea but I never knew which was what.

An important aspect of informal science approaches, as Alice described it, is the fact that they provide the means to make connections between school science and everyday life:

> I really enjoyed the fact that we did not go to university that day but had our class at the park instead! I had never had such an experience throughout my education. A visit to a park is something we usually do with our families, not with our instructors! Teaching science outside the classroom is probably the most constructive approach to teaching science. It has so many advantages: it provides a real-place for studying nature, it's about real life, it's interesting and fun because it's practical . . . it's not theoretical knowledge that students have to memorize from a book.

In her interview, Alice described how these informal science experiences impacted her views about science teaching and hence her identity as a science teacher:

> These experiences changed my views about science teaching and learning because I experienced science in ways that I never imagined. Throughout my schooling years I never had any informal science experiences except for a visit to the Science Museum in London with my parents. I believe that these settings offer great advantages because they make science learning fun and encourage students to appreciate science and the work of scientists . . . they kind of de-mystify science, making it more humane.

As indicated through Alice's narrative, she had been a low-motive science learner until she went to university. The figured world of her family did not offer her opportunities to experience science. As a young girl who grew up in the city, she had no experiences in nature. Science was one of her least favorite subjects at school. She articulated negative memories of learning science at elementary and high school when describing teacher-centered approaches, and this illustrates the impact of the figured world of traditional teaching on her identity work. Going to university, Alice's identity work became positively reinforced by various informal science experiences: an outdoor environmental study in collaboration with informal science educators, a visit to a botanical garden and a natural history museum, a collaboration with a scientist to conduct a study about snakes, and a visit to a science fair. These informal science experiences shifted her identity-work trajectory, which illustrates the influence of the university figured world. Alice envisioned herself as a teacher who stresses the relevance of science to society and hands-on activities and who highly values the role of informal science approaches in science teaching, revealing how she aligned herself to the figured world of science.

FUTURE DIRECTIONS

The cases of Maria and Alice illustrate how their science identities have been in formation from the early years of their lives and how various informal science experiences and interactions have shaped their identity work for science teaching. Even though there are profound differences in their identity works, commonalities exist regarding the impact of informal science experiences on their identity works. For example, throughout their schooling years both had negative experiences with science, associated with teacher-centered approaches and strict teachers. These findings speak directly to the need for research that examines beginning elementary teachers' identities as they enter their preparation program to better address their needs. It is crucial to examine beginning teachers' life histories in relation to science through the use of biographical approaches and to identify the pathways by means of which they come to develop as learners of science.

As evident in the findings of this study, both Maria and Alice could positively recall specific informal science experiences. In addition, it becomes clear in their narratives how specific informal science experiences they had during their teacher preparation impacted their identity work, specifically their views about science teaching and learning and the role of informal science approaches to science teaching. Both of them highlighted specific advantages that informal science contexts have to offer, such as having the potential to make science meaningful, fun, engaging, illustrating the value of science to society, and exemplifying the nature of science and the work of scientists.

These findings provide useful and important insights into how specific informal science experiences throughout their lives and within various informal and formal contexts impacted their identity work for science teaching. The findings provide a set of important implications for future research and teacher preparation, as they call for an examination of the role of informal science environments in science education. Several questions associated with the impact of informal science approaches and university informal science institutions' partnerships on the development of beginning teachers' identities for science teaching still remain unanswered. Some critical questions to examine are the following:

- In what ways could informal science settings be used in elementary teacher preparation?
- What examples of good practices in museum-university partnerships at the teacher preparation level does existing literature offer?
- In what ways could informal learning settings be used to support beginning teachers' development of reform-based identities for science teaching?
- In what ways do informal learning programs support beginning teachers' development of an understanding for the practice of science and the work of scientists?
- In what ways could informal learning environments be used to address reform recommendations in science education?
- What kinds of informal science experiences do beginning teachers have throughout their lives, and how do those impact their identity development?

REFERENCES

Avraamidou, L. (2014a). Developing a reform-minded science teaching identity: The role of informal science environments. *Journal of Science Teacher Education, 25,* 823–843.

Avraamidou, L. (2014b). Studying science teacher identity: Current insights and future research directions. *Studies in Science Education, 50,* 145–179.

Avraamidou, L. (2015). Reconceptualizing elementary teacher preparation: A case for informal science education. *International Journal of Science Education, 37,* 108–135.

Beijaard, D., Meijer, P., & Verloop, N. (2004). Reconsidering research on teachers' professional identity. *Teaching and Teacher Education, 20,* 107–128.

Carlone, H. B., Scott, C. M., & Lowder, C. (2014). Becoming (less) scientific: A longitudinal study of students' identity work from elementary to middle school science. *Journal of Research in Science Teaching, 51,* 836–869.

Holland, D., Lachicotte, W., Skinner, D., & Cain, C. (1998). *Identity and agency in cultural worlds.* Cambridge, MA: Harvard University Press.

Lave, J., & Wenger, E. (1991). *Situated learning: Legitimate peripheral participation.* Cambridge, MA: Cambridge University Press.

Varelas, M. (Ed.). (2012). *Identity construction and science education research.* Rotterdam: Sense Publishers.

Wenger, E. (1998). *Communities of practice. Learning, meaning, and identity.* Cambridge, MA: Cambridge University Press.

Epilogue
Looking Back to Look Forward

Wolff-Michael Roth & Lucy Avraamidou

This book is designed to bring together educators' writ large concerns with the learning of science in a variety of settings. Across the chapters, many authors reflect an often-observed consensus in characterizing certain settings as affording "informal science learning," including science museums, outdoor education programs, after-school science clubs, science exhibits, or using community-based resources in the teaching of science and teacher training and professional development. There are discordant voices, however, which challenge the very characterization of settings in terms of the adjectives formal and informal. Thus Roth (chapter 1), Dillon (chapter 5), and Dawson (chapter 7) take these characterizations to their limits, even suggesting that they are not so useful after all. On the other hand, Achiam and Nielsen argue for an emphasis on the science content when considering the purpose of informal science institutions.

CHALLENGING CONVENTIONS

In chapter 5, Dillon justly affirms our own position that (science) museums in a strong sense are not "informal" science learning places; Achiam and Nielsen (chapter 3) argue that science content needs to be at the core of the mission of out-of-school centers and museums; and even Rahm's "science clubs" are oriented toward science, thereby placing it into the foreground of everything else that might be going on. As Dawson (chapter 7) shows, even so-called informal science education opportunities may lead to negative experiences and dislikes. The exclusion of her research participants was apparent in exhibit texts and evident in the photographs displayed. There have been other reports about students' school-organized participation in out-of-school workshops run by environmental activists, where what was intended as fun was not at all experienced as fun (Roth, van Eijck, Reis, & Hsu, 2008). Why might there be such affectively negative forms of experience in settings thought about and organized as "informal science learning"?

Transgressing the Formal | Informal Dichotomy

A number of chapters focus on or include museums and science centers (chapters 2–5, 12, and 13). These are places separate from school but nevertheless characterized by formal structure and by their content. In the lives of students, museums, as schools, are only part of a plenitude of places in which they find themselves every day. All these places have some form of organization, as cultural-historical activity theorists readily point out. In all activities, human beings learn—although only some of them are formally denoted by the terms "education" or "science." Thus, for example, we do not tend to think of a welder working on an assembly line doing some spot welding as learning, just as we do not think of our dentist as learning while repairing a tooth or placing a crown. Yet we readily accept that practitioners get better at what they do by doing it day in and day out. The welder learns a lot about melting points, especially when finding out about joints that do not hold up, and the dentist gets better at understanding the physical and chemical properties of the different cements she is working with in the attempt to make a gold crown cling to the remainder of the tooth. There are, therefore, activities where "education" or "science" are made thematic and that therefore are characterized by institutional structures that orient to providing *learning* opportunities and, whether in-school, after-school, or out-of-school, are formal. On the other hand, there are other activities and places where people learn (science). It might have been a better alternative to choose a title for our book that reflects the linkages between schools and museums or other settings, each operating according to formal principles with the purpose of engaging people in *science*.

But thinking from within a disciplinary perspective, from within an institutional frame of science to theorize scientific literacy may be an even bigger problem. This has led us to suggest choosing a whole-life perspective and of science as but a strand among strands (Figure E.1). Such an approach makes sense when we think about life in general, outside of schools and science museums, where we do not tend to encounter science in a purified sense. As Figure E.1 shows, fibers are caught up in the strand; and the contribution of the fiber to the (tensile) strength of the strand is conditioned upon the existence of the strand as a whole, where other fibers have different properties. Not every strand has to be the same color (i.e., science). It is the whole strand that shapes, gives place to, and draws on the properties of the individual fibers. Although we may zoom in and focus on an individual fiber, we cannot understand its properties and contributions to the whole unless we think of it by means of a part–whole approach. Isolating science may not get us very far in our endeavor to move beyond dichotomies. Isolating formal educational learning from learning in all other settings that characterize our lives also may not get us any further.

In chapter 1, Roth describes the engagement of students in community activism, which is not focused on science as such, formal or informal.

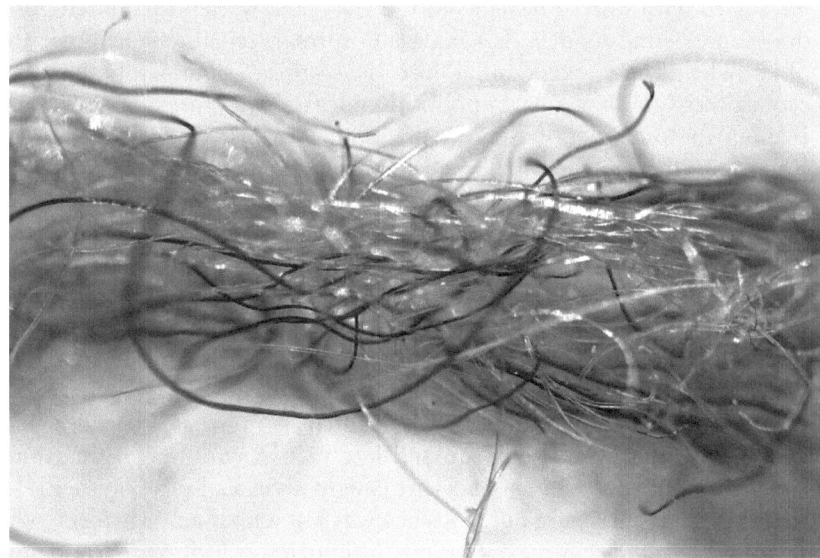

Figure E.1   Science is but a fiber in the strand that is the life of society, which is the condition for the existence of the former.

Instead, it is focused on watershed and community health. Among the many fibers that play into this focus, we may find science. But this does not have to be, and, as Roth's case shows, there are forms of engagement in reporting concerns for the health of the creek that require other things, such as communicating with aboriginal elders, operating a camera for visual documentation, recording interviews, and artistically organizing materials to make presentations that are rhetorically powerful. In all of these types of performances, we can rely on the division of labor that makes human culture so powerful so that we do not need any science that may lie behind (e.g., the physics of a camera) but instead focus on what is essential—e.g., photographs that are convincing evidence of the desolate state of a local creek.

Roth and his graduate student researched other places where science was but a strand among strands, and it was not unequivocal at all (e.g., Roth et al., 2004). Thus, for example, there were not only contradictions between the scientific assessments of the contamination of drinking water wells but also the parties with different stakes tended to mobilize those scientific reports and solutions that were in their own interest—such as a renowned political scientist and environmental lawyer, who advocated solutions to the water problems of residents unconnected to the village's water supply that favored a company in which he held a considerable number of stocks. Much as zooming and focusing on a single fiber eliminates a lot of information, the town engineers tended to discard some scientific and medical reports; and they also disregarded the (anecdotal) knowledge of the local residents, who have lived with water problems for 30–40 years. Rather than attempting to draw on existing models that integrate (formal) scientific and other forms of knowledge (e.g., experiential, historical, ethical, political, humanitarian, ecojustice-related), the early political endeavor was to exclude everything else but the one scientific report that spoke against extending an existing water line to cover the only 35 or so properties in a community of 15,000 souls that were not connected to the grid. How might those directly concerned, their advocates, and supporters engage with the kind of scientists with relevant knowledge? Not just any knowledge does the trick, as the residents saw, one of whom was a trained chemist. What is more important is a communicative competence and wisdom to communicate with scientists so that scientific literacy becomes an emergent feature suited to deal with the specifics of the problem at hand (Roth, 2003). In such cases, the formal | informal dichotomy no longer makes any sense, because science and engineering were caught up and intertwined with many other human endeavors—like the woolen fibers are caught up in a strand (Figure E.1)—so that they no longer were pure. Science was but a fiber among fibers, wound up into the strand that makes society. Although it is legitimate to isolate science or learning in formal educational settings from everything else for the purpose of research, we never reach any appropriate understanding unless we take into account the role of the fiber in the strand and the role of other fibers and the strand as a whole in the constitution of the function each fiber has in the whole.

Getting involved and contributing to such issues may be a different way of engaging students, thereby deinstitutionalizing—or, as chapter 1 suggests, deterritorializing—science education altogether. The presentation that the student Michelle in chapter 1 does denounces the practices of some people in her municipality, who dump garbage into the creek. Another student of the same age (12–13 years) publicly denounced two farms where the coliform counts in the creek dramatically increased. To make such denouncements stick, however, not just any claim can be made. Thus in the case of the two farms, the student had sought the assistance of a scientific laboratory to be able to establish coliform counts much more accurately than his middle school science laboratory allowed. A lot of science learning occurred—but not for the purpose of learning some scientific fact. Instead, the student was interested in making sure that his charge would stick. That is, he was concerned with the legal, ethical, and political clout that his endeavor would have. He ultimately reported the findings in an open-house event that a local environmental activist group had organized.

Transgressing the Boundaries of Boundary Conceptualizations

A frequent point relating to the distinction between formal and informal science learning is that of boundaries and boundary crossing (e.g., Bevan, chapter 4; Adams, chapter 15; Kisiel, chapter 16). But such discourse may actually be grounded more in theory than in everyday practice. For example, we easily move from our breakfast table to the grocery store, to work at the university, from where we take a break for a visit to the doctor or dentist; before finally returning home, we may stop by the center offering a class in Tai Chi. As we move across these activities that make our daily lives, there is a continuity of experience (Dewey, 1934/2008) that explodes any notion of boundary—even though we do very different things for very different purposes in these different places, drawing on our familiarities with the worlds that surround us. We feel ourselves even when we find ourselves out of place. In fact, to feel out of place requires the continuity with the place or places where we feel in place and at home. Why might it not make sense to think about science learning in the form of the boundary discourse?

For one, there is the human experience of continuity. Science educators sometimes theorize this continuity in terms of identity (e.g., Dawson, chapter 7; Katz, chapter 9; Avraamidou, chapter 17). On the other hand, sociologists, concerned with human beings in their lives more generally, have noted that identity, because of the different ways in which they are participating in different productive human activities, is fractured or fragmented. This situation had not gone unnoticed in practice theories and the cultural-historical activity theory based on them. Following Marx/Engels, the ensemble of social relations an individual lives and has lived is *personality*, a unit of analysis and category. As a result, there are as many aspects to personality as the activities in which a human being participates—including schooling, visits to museums, outdoor education, and science clubs. Although two individuals could potentially participate in exactly the same activities, which therefore constitute a common core to their experiences and personalities, their participation will neither be exactly the same nor will the relations between the motives of the different activities. These differences in the hierarchical relations between the motives of the different activities that we are and have been participating in constitute the differences. Personality, then, reflects the heterogeneous, *non-self-identical* nature of the person. Students who are confronted in school science with very different discourses than at home—which science educators have theorized in terms of the conflicts between naïve and scientific conceptions or between indigenous and Western science—do not somehow need to construct a third space (world). Instead, they experience themselves as living across and, thereby, integrating the two cultures.

Such integration can be observed among the most advanced scientists, who marvel at a beautiful sun*rise* or sun*set*, thereby topicalizing a sun that crosses the sky all the while talking about the rotation of the earth with respect to the sun when they are at work or teaching a course in astronomy. Those everyday experiences—some might be tempted to say those informal experiences with phenomena that science declares as its own—constitute the very condition for science to exist. Only with respect to an earth that does not move can any movement be experienced. A genetic approach to science education acknowledges the constitutive nature of students' everyday experiences to any science that may emerge; and this is so even though the latter may overturn what and how students previously have known. Visits to the museum, science center, participation in a discussion at a science cafe, carrying out a scientific experiment at a pop-up science shop, or outdoor center and participation in the science club, after-school science enrichment, or grassroots community group all provide opportunities for people to participate in activities organized around interesting phenomena. The form of participation is characterized by past experiences—Dewey (1934/2008) uses the analogy of a stone rolling down the hill to assist us in thinking about experience—and the extent of knowing one's way around the world. These forms of participation constitute the springboards for further engagement—just as every movement of Dewey's rolling stone depends on where and how it has moved before. For some persons, such engagement and the associated experience might be a starting point to get involved in science more deeply.

(Traditional) Schooling is the Problem

One of us repeatedly has advanced the idea that schooling is the root problem rather than the solution to any lack of interest in science (Roth, 2015). Short of the deinstitutionalization that Roth advocates, Dillon (chapter 5) offers an example of how schools might be organized differently to afford forms of engagement that differ from traditional schooling. He introduces Langley Academy, a school committed to holistic education that focuses on *museum learning*. Among others, as the school brochure describes, their Year 9 gifted and talented students co-curated a project with a science museum. Every month, students go on trips to museums and cultural heritage sites. However, as long as grades and grade reports are what schools produce, the motive of the activity has not changed and museum learning—or any one of the various forms the contributing authors describe in this book—can easily become just another way of doing the same. What if school science were entirely organized like a grassroots community group (e.g., Dawson, chapter 7), and after-school enrichment program or science club (e.g., Rahm, chapter 6)? In all of such settings, there are no grades and the motive of the activity is something other than receiving grades and accumulating symbolic capital for future academic progress.

One way to begin thinking about changing schools is to follow the example of Langley Academy. Thus why not organize school according to the principles of after-school (science) enrichment, outdoor education, or grassroots community groups? There would be many opportunities for increasing levels of participation not only in science but also in societal activities more broadly if elementary school students were running a community garden and donating or selling the produce they grow—such as the schools do participating in the School Grown project?[1] Why not have more schools commit to maintaining the viability of a salmon run in a local river by hatching salmon (which some students in British Columbia already do, see Roth, 2002)? Establishing and maintaining green corridors, turning wasteland into city gardens, or maintaining butterfly and other insect populations constitute further ways in which students may contribute to society and, in the process, learn. A colleague (J. Désautels) told us about how a biology teacher he had worked with had his students develop materials for educating young women in their community about reproductive health and then educated such women using the materials that they had produced. All such participation explodes the formal | informal dichotomy and provides new ways of thinking / theorizing schooling generally and school science more specifically. The principal purpose of schooling no longer exists in grades and grade reports but in doing/producing something in/for the community; that is, the principal purpose is to actively contribute to the life of society. The fundamental idea that underlies such conceptions is (a) the production of culture and cultural products and artifacts rather than their consumption or (b) the design and production of curriculum materials rather than their consumption, which, as studies have shown, leads to tremendous learning.

A focus on design and production of culture and cultural artifacts means— as currently organized only for the "gifted and talented" students at Langley Academy—that students could work on producing museum exhibits rather than consuming them, produce pamphlets and other materials to be used in outdoors centers, or organize rather than being consumers in (science) clubs. This can be done at all ages, such as seen in the exhibits some of our elementary (9–10 years of age) built in 1993 for an exhibit at Science World (Vancouver, British Columbia) devoted to bridges and bridge building or in the exhibits middle school students between 1998 and 2000 in Central Saanich (British Columbia) produced for the annual environmentalist open house where they also participated by talking to and educating visitors.

CODA

The chapters of this book constitute a beginning for rethinking science education. Even though the distinction between formal and informal may not hold up in the future as a useful way to think about education, it provides us with a starting point for ways in which schools and universities may be reorganized. Collectively, the chapters of this volume are grounded in the assumption that to make meaningful and transformative changes in science education we need to consider the characteristics of an era of globalization. Such characteristics are, for example, the diverse culture and racial origins of students, and language barriers caused by dramatic social changes happening in the world, such as border crossing and migration. These social changes call for a re-visioning of science education, which involves a conceptualization of science that goes beyond the binary oppositions of "formal" and "informal" science, and that moves beyond borders and boundaries, given that borders are limiting and considering that the boundaries of science are constantly shifting. In doing so, we ought to acknowledge that learning environments are nowadays hybrid and multicontextual and that learning experiences are complex and multidimensional. The contributors to this volume problematize various aspects of formal science education and propose new ways of conceptualizing science education shaped at the intersection of personal, local, and global realities. At the heart of the account of this proposition is an examination of the science learning that takes place beyond the formal school realm, in-between spaces, as the authors question the taken-for-granted spaces, boundaries of science, and traditional discourses. Theoretically grounded within these understandings, the authors provide examples of authentic and unique opportunities offered in diverse settings, for teachers and students to empower themselves to experience science in meaningful, exciting, and culturally relevant ways.

If whatever the contributing authors think about or practice as informal science were to become the organizational pattern for school science learning, we would already have made a first step. As indicated in the preceding sections, the proof of concept for making such steps has already been provided in special projects around the world. The question now is not how to "scale up" any such approach but to engage schools in developing their own approaches within their unique contexts in a situationally appropriate manner that does take into account their resources as well as the local community characteristics. Thus the student involvement in international service projects—e.g., the International Service Learning Program at Appleby College, Oakville, Canada[2]—may be feasible only in a private school, however, maintaining a market garden and selling produce is possible even in areas serving impoverished populations. As A. Barton has shown through her work in New York, students living with their families in homeless shelters may learn while turning an abandoned city lot into a neighborhood park. Education of future generations is much too important to leave it to schooling traditional conceived. The *informal* aspect and side of our collective considerations in this book constitutes a great way to rethink what we might offer students at all ages, including those who are attending college and university programs to become teachers themselves.

NOTES

1 See, for example, http://www.foodshare.net/schoolgrown or http://www.theguard ian.com/environment/2013/apr/02/waitrose-alan-titchmarsh-schools-initiative.
2 http://www.appleby.on.ca/page.cfm?p=2699

REFERENCES

Dewey, J. (1934/2008). *Later works vol. 10: Art as experience* (J.-A. Boydston, Ed.). Carbondale, IL: Southern Illinois University Press. (First published in 1934)

Roth, W.-M. (2002). Taking science education beyond schooling. *Canadian Journal of Science, Mathematics, and Technology Education, 2*, 37–48.

Roth, W.-M. (2003). Scientific literacy as an emergent feature of human practice. *Journal of Curriculum Studies, 35*, 9–24.

Roth, W.-M. (2015). Schooling is the problem: A plaidoyer for its deinstitutionalization. *Canadian Journal for Science, Mathematics, and Technology Education, 15*, 315–331

Roth, W.-M., Riecken, J., Pozzer, L. L., McMillan, R., Storr, B., Tait, D., . . . Pauluth Penner, T. (2004). Those who get hurt aren't always being heard: Scientist-resident interactions over community water. *Science, Technology, & Human Values, 29*, 153–183.

Roth, W.-M., van Eijck, M., Reis, G., & Hsu, P.-L. (2008). *Authentic science revisited: In praise of diversity, heterogeneity, hybridity*. Rotterdam, The Netherlands: Sense Publishers.

Index